Science Now 5

Revised Edition

Series Editor
Jyothsna Natraj

Collins

Science Now 5

Revised Edition

Series Editor
Jyothsna Natraj

Published by Collins Learning
A division of HarperCollins*Publishers* India Private Limited
HarperCollins *Publishers*
A-75, Sector 57, Noida, Uttar Pradesh 201301, India
1 London Bridge Street, London SE1 9GF, United Kingdom
2 Bloor Street East, Toronto, Ontario M4W 1A8, Canada
Lvl 13, 201 Elizabeth Street (PO Box A565, NSW, 1235),Sydney NSW 2000, Australia
195 Broadway, New York, NY 10007, USA

Browse the complete Collins catalogue at **www.collins.in**
for any queries/feedback, please write to us at
SCHOOLSUPPORT@harpercollins.co.in
or call us on **0120-4044800**.

First edition 2015
Latest revised edition 2018

Reprint: 12

ISBN: 978-93-5277-197-4

Content developed by
EXCELSOFT TECHNOLOGIES PVT. LTD.

Artwork and layout by **AptaraCorp Ltd., Mammoth Studioz Ltd.**

Image research by
EXCELSOFT TECHNOLOGIES PVT. LTD.

Printed and bound in India by
Thomson Press India Ltd.

Acknowledgements

Cover: GMVozd, smarques774

Title Page: GMVozd, smarques774

Inside: **Page 10**: Przemek Klos, Jerry Horbert, Martin Fowler, Ethan Daniels, Jeanette Dietl, chinahbzyg, Martin Fowler, MVPhoto; **11**: Oksana Alekseeva, aida ricciardiello; **13**: ChaiyonS021, David Lade; **18**: Jerry Horbert, Oksana Alekseeva, MVPhoto, chinahbzyg; **80**: Excelsoft; **84**: eddtoro, aletermi; **103**: Tom Wang, fotostory, Cico; **105**: Igor Zh.; **106**: stockphoto mania; **107**: vallefrias; **111**: Tom Wang, Catmando, stockphoto mania, Igor Zh.; **112**: Simon Krzic, Mr Doomits, Catmando; **113**: Jiang Zhongyan, Siim Sepp, Siim Sepp; **114**: sonsam, J.Schelkle, Siim Sepp, Siim Sepp, nito, Siim Sepp, Alaettin YILDIRIM, Tyler Boyes; **115**: teena137, hxdbzxy,
idea for life, anat chant, Imfoto; **120**: Excelsoft, Excelsoft, Siim Sepp, sonsam, Tyler Boyes; **132**: Andrey_Kuzmin, konzeptm; **136**: Andrey_ Kuzmin, konzeptm, **160:** Andrey_Kuzmin

Preface

***Science Now* (Revised Edition)** is a series of eight books for Classes 1 to 8. This series conforms to the vision of the *National Curriculum Framework*.

Based on the NCERT syllabus, the series follows a coherent and structured approach, which provides a seamless continuity in the science curriculum for classes 1 to 8, and even thereafter.

This revised edition retains the focus of the original, where development of lateral thinking, analytical, research and deduction skills in students were encouraged.

The series has been revised taking into account the extensive feedback from user schools and experienced teachers. The content has been updated and the language simplified to cater to the needs of all kinds of learners in a classroom.

Some Key Features

Let's Get Started to revise the concepts learnt earlier

Science in Context gives interesting extra information on the topic at hand

Watch out! busts some common myths or warns against harmful habits/actions

Hands-on **activities** to hone experimental and deduction skills

Go Green gives environment-friendly tips and advice

Summary gives a snapshot of the chapter for quick recapitulation

Higher Order Thinking Skills (HOTS) challenges to think beyond the textual knowledge

Cross-curricular Learning Links (for classes 1 to 5) integrate knowledge across subjects

Life Skills has questions testing awareness, empathy and attitude towards self and others

Project questions to link the concept learnt to real-life applications

Pictorial **worksheets** at the end of each chapter

In the Laboratory (for classes 6 to 8) tests skills for experimental investigation, observation and deduction

In the Library (for classes 6 to 8) gives topics for further reading and research

Latest **National Science Olympiad** paper to help students prepare for competitive examinations

Four **assessment papers** and two **comprehensive assessment papers** have been given at the end of each book, in addition to the exercises at the end of each chapter.

In line with the CBSE guidelines, features for evaluation and tools of assessment have been provided extensively to the teachers and learners in a well-integrated manner.

Feedback, valuable comments and suggestions from the users are welcome.

Authors

Key Features

Science in Context

A flying insect called firefly gives out light from its tail that shines in the dark.

Watch Out!

Poisonous snakes have sharp curved teeth called fangs. These fangs are used to inject poison (called venom) into the prey.

Go Green

With the help of an elder, grow some plants in your garden at home or school. Water them regularly.

Word Web

[1]**Canvas:** a strong, thick cloth

[2]**Gypsies:** people who roam around from place to place

G HOTS (Higher Order Thinking Skills)

Why does a balloon grow bigger in size when air is blown into it?

(Hint: Does air take up space?)

Fun Activity (Cross-curricular)

A list of words related to plants is given below. Circle the correct spelling from each group.

1. Hrebs	Herbs	Harbs
2. Tunk	Trunk	Trink
3. Stems	Stums	Steam

I Life Skills

Look at the pictures and answer the question given below.

J Project

Make a pinwheel.

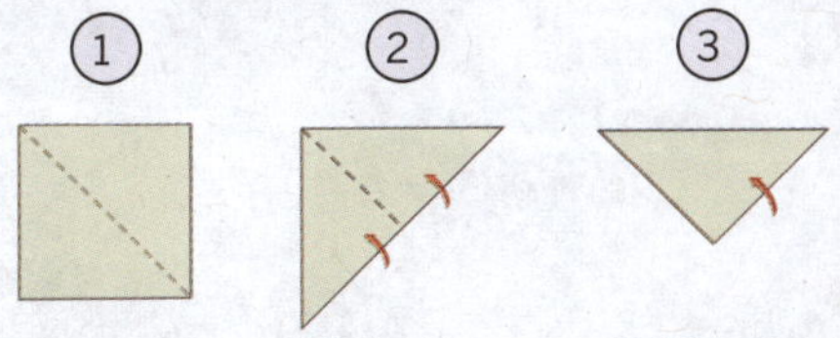

1. Take a square paper.
2. Fold the paper diagonally.
3. Fold once again as shown in the picture.

Summary

1. Moving air is called wind.
2. Wind helps us in many ways.

WORKSHEET

1. Draw arrows to complete the water cycle and label the diagram.

Contents

1 Growing Plants

Learning Checklist

- New plants from seeds, roots, stems, leaves and spores
- Agriculture

Let's Get Started

Look at the pictures of the fruits given below. Write whether each of these fruits have no seed, one seed, few seeds or many seeds.

____

____

____

____

____

____

New plants mostly grow from seeds. So, we can say that seeds help the plant to reproduce. But there are also some plants that do not produce seeds. How do they reproduce then?

In this chapter, let us learn about the various ways in which plants reproduce.

NEW PLANTS FROM SEEDS

Most of the fruits and vegetables have seeds. New plants grow from these seeds. Let us do an activity to understand the structure of a seed.

Activity 1

Aim: To remove the seed coat and observe the structures inside a seed

Materials required: Some kidney beans *(rajma)*, water and a plate

Procedure:

1. Soak somc kidney bean seeds overnight in water.
2. The next day, you will see that the seeds have absorbed water and are swollen.
3. Remove the outer covering of the seeds.

This outer covering is called the seed coat. Now let us study the structure inside the seed.

The seed coat protects the seed and the baby plant inside it. There is a tiny opening in the seed coat called the **micropyle** through which the seed absorbs water. Above the **micropyle** is a scar called **hilum**, which shows the place where seed was attached to the fruit. The structure inside the seed coat can be broken down into two halves. Each half is called a **cotyledon** or **seed leaf**. The seed leaves store food for the developing baby plant, also called an **embryo**. Seeds such as peas, pulses and beans have two cotyledons. Seeds such as maize, rice and wheat have only one cotyledon.

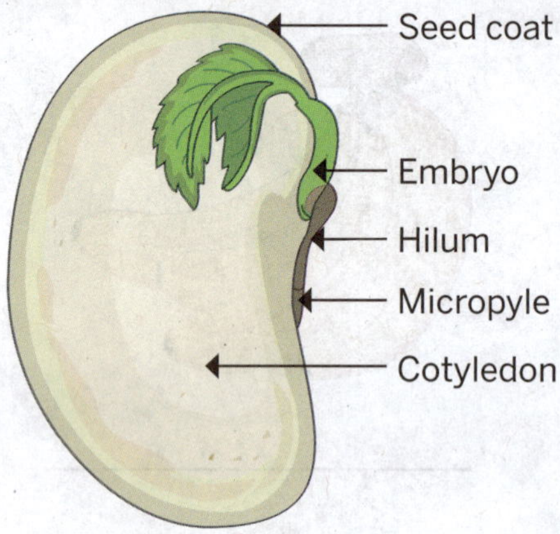

Parts of a seed

Not all seeds grow into new plants. Some seeds are destroyed by harsh weather conditions such as strong winds and heavy rain, while some are eaten by birds and animals as food. A seed needs sufficient amount of water, light, warmth and air to grow into a new plant. The process in which a new plant grows from a seed is called **germination**.

1. A seed sown in soil gets warmth, air, light and water.
2. The seed coat breaks and a baby plant emerges. It grows small roots.
3. The roots grow further and a small shoot develops.
4. As the roots and the shoot grow in size, leaves appear, and the cotyledons shrink and eventually fall off.

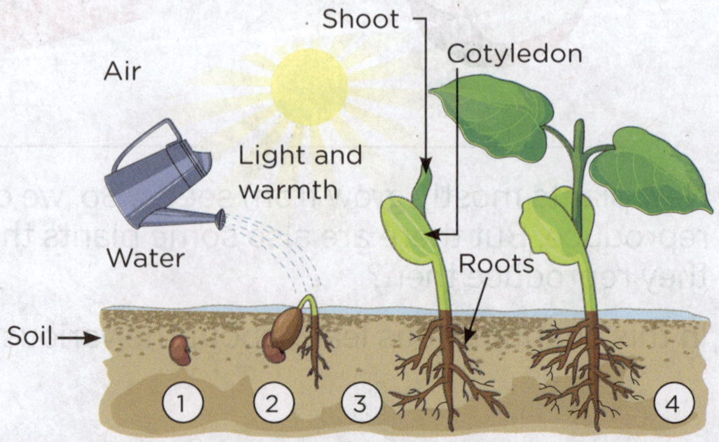

Stages of germination in a seed

Activity 2

Aim: To study the conditions necessary for the germination of seeds

Materials required: Bean seeds, a bowl, three plastic containers, cotton cloth, foil paper and water

Procedure:

Set-up 1:

- Soak a few bean seeds in a bowl of water for about 30 minutes.
- Dampen a cotton cloth with water and place it in a plastic container.
- Place the soaked seeds in the plastic container and cover it with foil paper.
- Then place the container in a dark place for two days, undisturbed.

Set-up 2:

- Take a handful of seeds in a plastic container and add excess amount of water to it such that the seeds float in water.
- Place the plastic container under sunlight/artificial light, undisturbed for two days.

Set-up 3:

- Take a handful of seeds in a plastic container and close it with a lid.
- Keep the plastic container near a window. Leave it undisturbed for two days.

Observation: Fill in the table based on your observations.

Experimental set-up	Conditions	Observation
Set-up 1		
Set-up 2		
Set-up 3		

Conclusion: ____________ and ____________ are necessary for the germination of seeds.

Dispersal of Seeds

If seeds are sown very close to each other, they will struggle for space, light, air, water and nutrients. Therefore, it is necessary that the seeds are scattered far away from each other. The process in which seeds are scattered far away from the parent plant to obtain suitable conditions for germination is called **dispersal of seeds**. In nature, seed dispersal is carried out by wind, water and animals. They are called agents of seed dispersal.

Dispersal by Wind: Some fruits have very light and hairy or wing-shaped seeds that get caught in the wind and are easily blown away to far-off places. These seeds germinate in fertile soil if they get favourable conditions.

Examples of plants whose seeds are dispersed by wind are dandelion, cotton and maple.

Dandelion

Cotton

Maple

Dispersal by Water: Seeds of some fruits have a waterproof covering and are so light that they can float on water and are carried to far-off places. For example, the coconut has a heavy seed inside, but it floats on water as the seed is covered by spongy layers of fibre filled with air. Other examples of plants whose seeds are dispersed with the help of water are lotus, water lily and foxglove.

Coconut

Water lily

Dispersal by Animals: Some seeds have hook-like structures that help them to cling on to animals' fur or have sticky substances that stick to animals' skin. This way, the seeds are carried away from the parent plant. Many birds and animals eat fruits along with their seeds. The seeds come out of their bodies in their droppings and are scattered to different places where they grow into new plants. Examples of plants whose seeds are dispersed by animals are cocklebur, mistletoe and berries.

Cocklebur

Mistletoe

Berries

Dispersal by Explosion: Some seeds have a leathery covering or are found inside pods that explode or split open when exposed to the sun or when they dry up or even when touched. When the pods burst open, the seeds are scattered far away from the parent plant. Examples of plants whose seeds are dispersed by explosion are pea, gorse, peanut, balsam and castor.

Peas

Gorse

Revise and recall

1. **Match the seeds with their corresponding dispersal method.**

Cotton	Water
Gorse	Wind
Coconut	Animal
Berries	Explosion

2. **Answer in one word.**
   a. The process in which a new plant grows from a seed ______
   b. A plant that has light and hairy seeds ______
   c. A plant whose seeds get dispersed by animals ______
   d. A plant that disperses its seeds by explosion ______

New plants can grow not only from seeds, but also from roots, stems, leaves and spores. Let us learn how.

NEW PLANTS FROM ROOTS

Roots of plants such as sweet potato, carrot and beetroot store food. These roots contain small buds[1] from which new plants grow. A piece of the root along with the bud is cut and planted in the soil. Under suitable conditions, these buds develop roots and shoot, forming a whole new plant.

Carrot

Beetroot

Word Web

[1]**Bud**: a small lump on the plant part from which new shoot, leaf, flower or a new plant can grow

NEW PLANTS FROM STEMS

Various natural and artificial methods can be used to obtain new plants from stems. Some of them are as follows.

Underground Stems: Some plants have stems that grow below the ground. Such stems are called underground stems, for example, potato, ginger and onion. These stems have small buds on them. In potatoes, these buds are called **eyes**. Any portion of a potato with an eye can grow into a new plant.

Stem Cutting: In plants such as rose, bougainvillea and money plant, a part of the stem containing a bud is cut and planted in the soil. It develops roots after some time and finally grows into a whole new plant. This method is called **stem cutting**.

Stem cutting

Grafting: In plants such as rose, mango and apple, a portion of the shoot is cut and then attached to the shoot of another plant. The part of the stem that is cut is called the **scion** and the part to which it is attached is called the **stock**. This method is called **grafting**.

Layering: In plants such as jasmine and raspberry, a branch of the plant is bent until it touches the ground. A portion of it is covered with soil, leaving the growing tip exposed. The portion covered with soil develops roots after a few days. It is then cut from the parent plant and grows as an independent plant. This method is called **layering**.

Plants such as strawberry also reproduce through layering as they have horizontal stems that grow near the ground. These stems are called **runners**. They form roots at regular intervals and produce new plants.

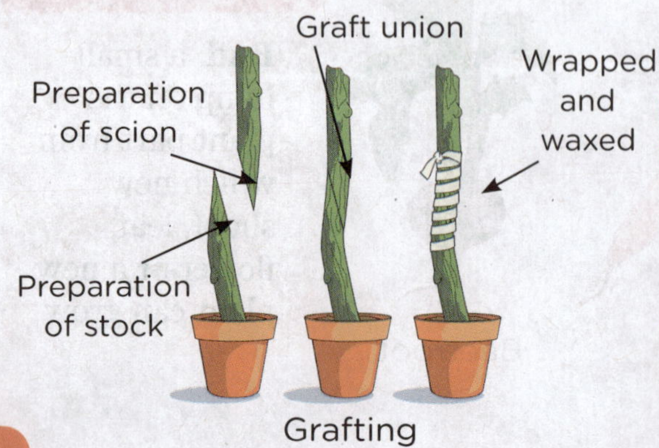

Grafting

Layering

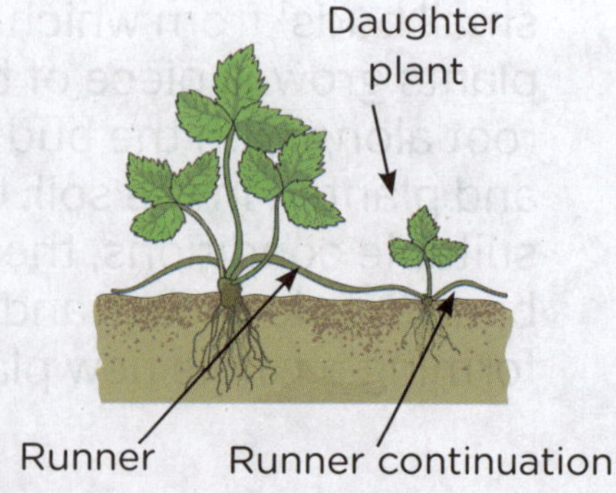

Runners

NEW PLANTS FROM LEAVES AND SPORES

In plants such as *Bryophyllum* and *Begonia*, new plantlets can grow from the well-developed leaves while these leaves are still attached to the branch. Once these plantlets develop, they drop off or detach themselves from the parent plant. They grow into new plants on moist soil.

Bryophyllum

Ferns

In some plants such as ferns and mosses, new plants grow from special structures called **spores**[1]. These are very light and tiny powdery structures. They can survive for a long period of time in unfavourable conditions.

Word Web

[1]**Spores**: tiny seeds that appear powdery when touched

Revise and recall

Fill in the blanks using the correct words.

1. Roots of __________ have buds from which new plants grow. (beetroot/rose)
2. __________ can grow into new plants when its stem is cut and planted in soil. (Money plant/Onions)
3. Horizontal stems in strawberry that form roots at regular intervals and produce new plants are called __________. (runners/scion)
4. __________ produce spores on their leaves which grow into new plants. (*Bryophyllum*/Ferns)

AGRICULTURE

A plant grown in large numbers on a plot of land during a particular season is called a **crop**.

The practice of growing crops on a large scale in fields is called **agriculture**.

Science in Context

Around one-third of the human population is involved in agriculture.

There are various processes involved in agriculture. They are as follows:

1. **Preparing the soil by ploughing the field**: Farmers use **bullocks** or **tractors** to plough their fields. Ploughing turns the soil and loosens it.

Ploughing using bullocks

Ploughing using tractor

Word Web

[1]**Weeds**: unwanted plants growing among crops

[2]**Pests**: insects and animals that harm the crops

2. **Adding manure or fertilizers to the soil**: When the farmers grow crops year after year, the soil becomes less fertile. So, the farmers add natural manure or chemical fertilizers to increase the fertility of the soil.

Manual sowing

Seed drill

3. **Sowing healthy, good-quality seeds**: Healthy seeds are sown either manually in grooves made by the farmers while preparing the field or through machines called seed drills.
4. **Irrigation (watering) of land**: Watering the land regularly for the crops to grow well is called **irrigation**. Modern methods of irrigation such as sprinkler are common nowadays.

Sprinkler

Spraying chemicals

5. **Protecting the land from weeds[1] and pests[2]**: The farmland is protected from weeds and pests by spraying chemicals called weedicides and pesticides, respectively.
6. **Harvesting**: The fully grown crops are cut and collected for use. This process is called harvesting. Harvesting can be done manually or with the help of machines called harvesters.
7. **Storing**: Harvested crops are stored in dry, well-ventilated places that are free from germs and pests.

Science in Context

To increase the fertility of soil, farmers often plant different variety of crops in a field in successive years. The nutrients used by one type of crop is restored by planting the other variety. This method is called crop rotation.

Manual harvesting

Using a harvester

Storage

Go Green

Growing plants in barren lands not only adds to the production of food to meet human needs, but also gives us pure air and prevents soil erosion.

Different crops need different climatic conditions to grow. Some crops grow well in hot climate while some grow well in cold climate; some need less water to grow while some need a lot of water and hence grow well in areas that receive heavy rainfalls. Therefore, farmers grow crops based on the climate suitable for each crop.

Depending on the time of the year or the season in which they are grown, crops are of two main types: kharif and rabi.

Kharif crops: Kharif crops are usually sown in the beginning of June and harvested in September–October. These crops are grown during this season because they need plenty of water for good yield, which they get from the monsoon rains.

Examples of kharif crops are cotton, maize, paddy, sunflower, soybean and millet.

Rabi crops: Rabi crops are grown in winter from November and harvested in April. These crops do not require much water for growth.

Examples of rabi crops are cauliflower, beans, peas, wheat, mustard, gram, barley and oats.

Let's Say Together

Germination	Hilum	Cotyledon	Dispersal
Irrigation	Harvesting	Kharif	Rabi

Keywords

Seed coat: The outer covering of a seed that protects the structure inside it
Cotyledon: The part of the seed that contains food for the baby plant
Embryo: The part of the seed from which a new plant grows under suitable conditions
Micropyle: A tiny opening in the seed coat
Dispersal: The process in which seeds are scattered far away from the parent plant to obtain suitable conditions for germination
Agriculture: The practice of growing crops on a large scale in fields
Harvesting: The process of cutting and collecting fully grown crops from fields for use
Kharif crops: Crops that are grown in the monsoon season, that is, from June to October
Rabi crops: Crops that are grown in the winter season, that is, from November to April

Summary

1. Plants reproduce by seeds or other plant parts such as roots, stems, leaves and spores.
2. A seed consists of a seed coat, hilum, embryo and cotyledons.
3. Seeds are scattered far away from the parent plant to obtain suitable conditions for germination.
4. The process in which a new plant grows from a seed is called germination.
5. The practice of growing crops on a large scale in fields is called agriculture.
6. Crops grow in different seasons and in different conditions.
7. Crops grown from June to October are called kharif crops.
8. Crops grown from November to April are called rabi crops.

EXERCISE

A Choose the correct option for each of the following.

1. The part of a seed from which a new plant grows is
   (a) Hilum (b) Cotyledon
   (c) Embryo (d) Seed coat
2. The seeds of ______________ plant is dispersed by wind.
   (a) Cocklebur (b) Berries
   (c) Mistletoe (d) Cotton
3. The plant that can reproduce by stem cutting is
   (a) Bougainvillea (b) *Bryophyllum*
   (c) Potato (d) Bean
4. Kharif crops are usually grown during the months of
   (a) June–August (b) June–October
   (c) November–April (d) November–January
5. Pea and wheat are ______________ crops.
   (a) rabi (b) kharif
   (c) both (d) none

B **Fill in the blanks using the correct words.**

1. ______________ seeds have one cotyledon. (Pea/Maize)
2. Seeds of ______________ are dispersed away from the parent plant when their pods burst. (maple/peas)
3. ______________ contain small buds called eyes from which new plants can grow. (Roses/Potatoes)
4. Tiny powdery structures seen on the leaves of ferns are called ______________. (hooks/spores)
5. A ______________ plant can be grown by layering. (rose/raspberry)

C **Match the following.**

Column A	Column B
(a) Moss	(1) Dispersal by water
(b) Ginger	(2) New plant from roots
(c) Foxglove	(3) New plant from spores
(d) Cocklebur	(4) Underground stem
(e) Beetroot	(5) Dispersal by animals

D **State whether the following statements are true (T) or false (F).**

1. Dispersal of seeds avoids overcrowding of plants in one place. ☐
2. Rose plant has plantlets on its leaves that grow into new plants. ☐
3. In onion, new plants can be grown from the stem. ☐
4. Irrigation is done before sowing the seeds in the soil. ☐
5. Rabi crops depend on the monsoon rains for good yield. ☐

E **Answer the following in one word.**

1. The process in which a new plant grows from a seed ______________
2. A dispersal agent for wing-shaped seeds ______________
3. The process in which a part of the stem containing a bud is cut and planted in the soil ______________
4. Plants grown in large numbers on a plot of land during a particular season ______________
5. The practice of growing crops on a large scale in fields ______________

F **Answer the following questions briefly.**

1. What is the function of seed coat?
2. What is germination?
3. What are the basic requirements for a seed to germinate?
4. What makes a coconut float on water although the seed inside it is heavy?
5. When does a fruit explode to disperse the seeds inside?
6. List three parts of plants other than seeds from which new plants can grow.
7. Name the two ploughing methods used by farmers.

G **Answer the following questions.**

1. What is seed dispersal? Discuss the various agents of seed dispersal with suitable examples.
2. With the help of a diagram, explain the process of stem cutting and grafting.
3. How are crops grouped? Give two examples for each type.
4. List the various steps involved in agriculture. Explain the processes of irrigation and harvesting.

H **HOTS (Higher Order Thinking Skills)**

1. What are the ways in which you can keep potted plants healthy?
(Hint: Conditions needed for plant germination and growth)
2. Riya had kept her potted plant in a dark corner of her room. Her mother asked her to keep it near a window or in the balcony. Can you tell why? Look at the picture given alongside for hint.

3. Why do farmers add manure to the soil?
(Hint: How does manure help the soil?)

I **Identify the plants in the following images and write down the method of seed dispersal that takes place in each of these plants.**

_______________ _______________ _______________ _______________

_______________ _______________ _______________ _______________

Label the given parts in the structure of a bean seed.

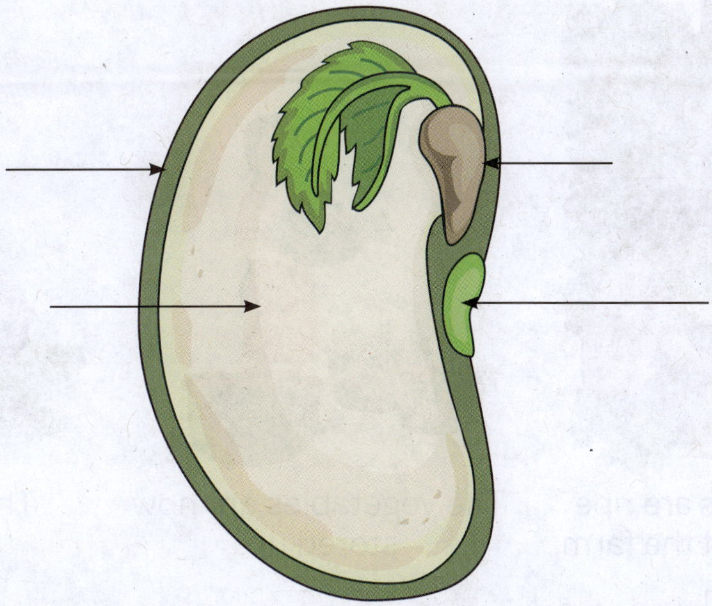

J Project

Soak 3 or 4 beans overnight in a bowl. Plant them in a pot the next day. Observe after two or three days. Record your observations. Continue to observe the bean plants till they produce their own beans. Keep recording your observations periodically. Take photographs of each stage, prepare a report and present it in your class.

K Cross-curricular Learning Link

Read newspaper articles about farmers and write your own article based on what you read and present it in your class.

L Life Skills

We can grow our own vegetables in a kitchen garden. Ask the elders of your family to make a small garden and grow a few vegetables. Kitchen wastes like vegetable peels, waste food, etc. can be used to make manure for the plants. Water the plants regularly and when the vegetables are fully grown, harvest them and enjoy them in a meal!

List six of your favourite vegetables that you would like to grow in your kitchen garden.

1. ____________________ 2. ____________________

3. ____________________ 4. ____________________

5. ____________________ 6. ____________________

WORKSHEET

Imagine you are going to grow your own vegetables in a vegetable farm. Observe the images given below and number them in proper sequence to show how you will grow the plants and take care of them.

The vegetables are ripe for harvesting at the farm.

☐

The vegetables are now stored up.

☐

The seeds are sown.

☐

The plants are irrigated by watering them regularly.

☐

Preparing the soil

☐

Adding weedicides and pesticides

☐

2 Animal Habitats and Adaptations

Learning Checklist

- Habitat
- Adaptation
- Migration

Let's Get Started

The pictures given below show different animals in their natural environment. Choose from the given words and write the names of these places in the space provided.

Forest Pond Desert Polar region

____

____

____

____

HABITAT

The environment where an organism naturally lives and reproduces is called its **habitat**. Different organisms live in different habitats. They obtain food, shelter, air and water from their habitat.

Let us learn about the different types of habitats found on the Earth.

Forest

Forest habitats usually have a large number of plants and trees. A variety of animals live in forests in various places such as caves (e.g. tiger and lion), on trees (e.g. birds and monkey), in burrows (e.g. snake and rabbit) and on the forest floor (e.g. deer, elephant and insects). Many animals that live in the forests have well-developed limbs[1] that help them to run and walk. Limbs also help them to hunt or run away from predators[2].

Forest habitat

Word Web

[1]**Limbs**: hands and legs

[2]**Predator**: an animal that kills and eats other animals

[3]**Scanty**: very little

Desert

Desert habitats are sandy lands with scanty[3] rainfall. Most deserts are very hot and dry during the day and become cold at night. Animals usually found in deserts are camel, kangaroo rat, rattlesnake, etc. Animals in deserts have special features to survive there. For example, camels have thick, broad feet that help them walk easily on loose sand. They also have thick eyelids that protect their eyes from the sand blowing with the wind. They store water and fat in their humps for use in times of scarcity. Some desert animals live in burrows to escape heat.

Desert habitat

Science in Context

The Gobi desert in Asia and the deserts in the continent of Antarctica are always cold. They are called cold deserts.

Polar Regions and Mountains

The high mountains and the Polar regions (the Artic and Antarctica) are covered with snow and are very cold throughout the year. During winter, the temperature can drop to very low degrees. Some animals found in the Polar regions are polar bear, penguin, seal and walrus. Animals like polar bears have thick fur on their body that keeps them warm. Penguins have a layer of fat under their skin called **blubber**. It helps in keeping their body warm.

Polar habitat

Ocean

Oceans are the largest habitats of the world and contain salt water. Pacific Ocean, Arctic Ocean, Atlantic Ocean, Southern Ocean and Indian Ocean are the various oceans on the Earth.

Big and small fish of different types are found in salt water habitats. Crabs, starfish, sharks, whales, prawns, sea turtles, sea horses, jellyfish, electric rays, octopuses and many other types of animals are found here.

Ocean habitat

Freshwater

Freshwater habitats include various ponds, lakes, streams and rivers. Different kinds of fish, snakes and frogs live in this habitat. Frogs use their long hind legs and webbed feet for swimming in water. Water birds such as duck and swan swim in water with their webbed feet. Cranes have long legs to wade through water.

Fish

Frog and duck

Cranes

Freshwater habitats

ADAPTATION

The features and habits of organisms that help them to survive in their habitats are called **adaptation**. Let us learn about some adaptations of animals living in different habitats.

Body Coverings

Animals have different types of body coverings such as thick fur, scales and shell. Let us learn about them.

Scales: Bodies of reptiles such as crocodile, lizard and snake are covered with scales. Snakes shed their skin periodically. The process of shedding of old skin to replace it with new skin is called **moulting**. The body of fish is also covered with scales.

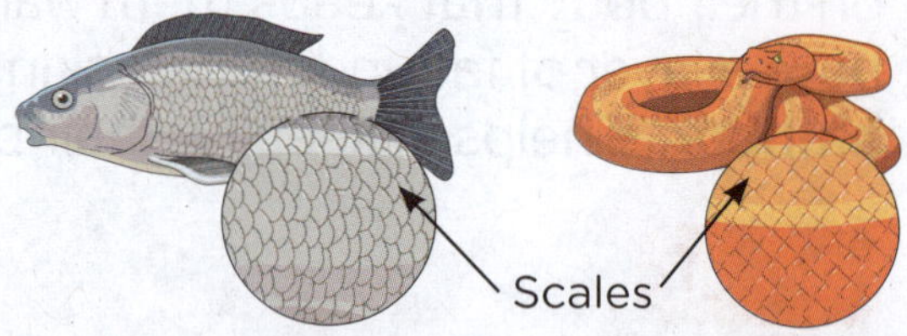

Shell: Shell protects the soft bodies of animals such as turtle, tortoise and snail from damage. The hard shell also protects them from being eaten by predators.

Fur or Wool: The body of animals such as polar bear and yak is covered with thick **fur**. A sheep's body is covered with **wool**. The fur/wool helps the animals to keep themselves warm in cold weather.

Yak

Sheep

Feathers: The body of birds is covered with feathers. Birds have three types of feathers:

Flight feathers are present on the wings and help the bird to fly.

Body feathers cover the body.

Down feathers keep the body warm.

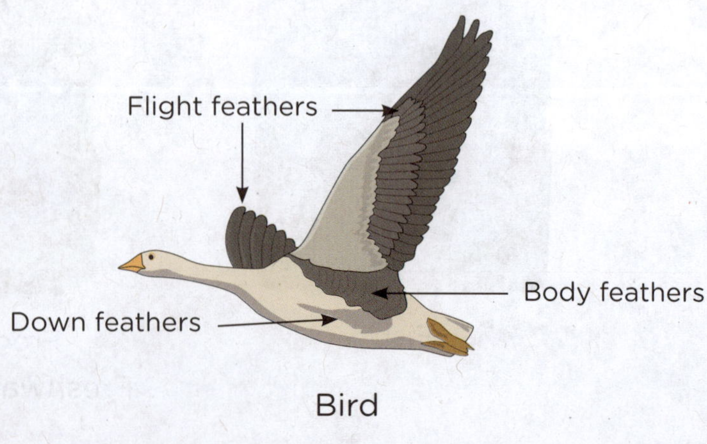

Bird

Hard body covering: Insects such as grasshopper and cockroach have a hard body covering made of a substance called **chitin**.

Cockroach

Revise and recall

Name the following.

1. The environment where organisms naturally live and reproduce __________
2. The body covering of a crocodile __________
3. The body covering of a snail __________
4. The feathers that help birds to fly __________

Breathing Organs

Different animals have different breathing organs suited to their habitats.

Earthworm breathes through skin

Skin: Animals such as earthworms use their moist skin for breathing. The exchange of gases takes place through the surface of their skin. A frog breathes through its skin when it is inside water.

Body surface: Small microscopic[1] organisms breathe through their body surface. Examples are *Amoeba* and *Paramoecium*.

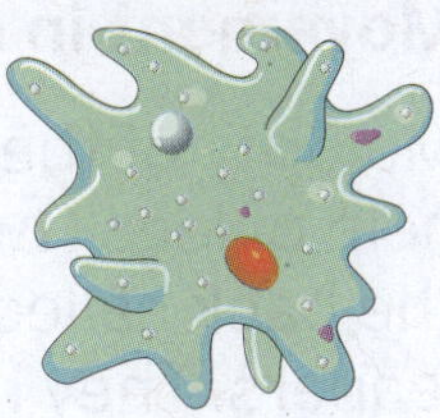

Amoeba breathes through body surface

Word Web

[1]**Microscopic**: not large enough to be seen with naked eye but visible under a microscope

Gills: Most aquatic animals such as fish and prawns breathe with the help of gills.

Water flows in through the mouth and flows out through the gills. The gills help in absorbing the oxygen dissolved in water.

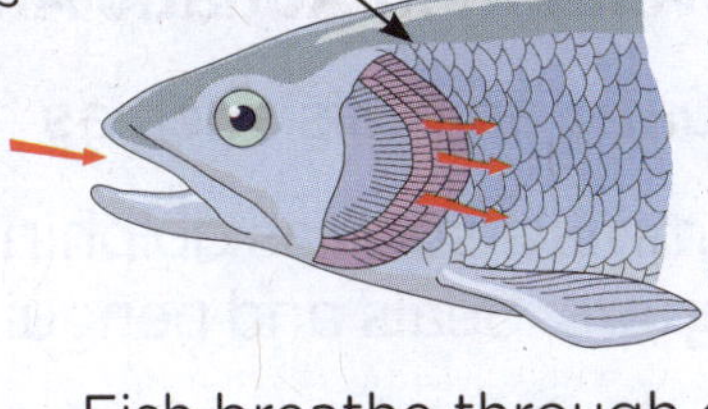

Fish breathe through gills

Spiracles: Insects have special structures for breathing known as **spiracles** (also called **air holes**). These are small openings present along the side of the insect's body.

Air enters the insect's body through these spiracles and is expelled out through the same structures.

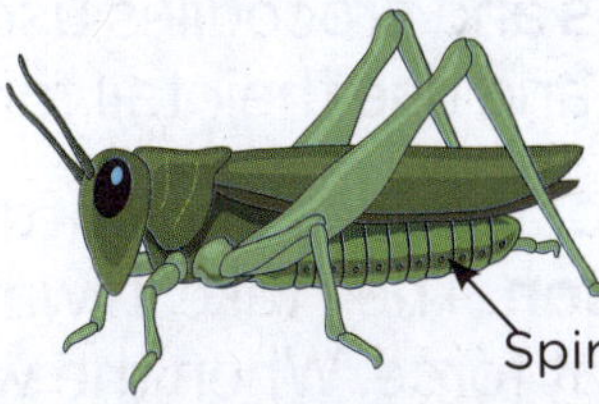

Insects breathe through spiracles

Science in Context

The blood of insects is white in colour as it does not contain **haemoglobin** (the pigment that gives the red colour to our blood).

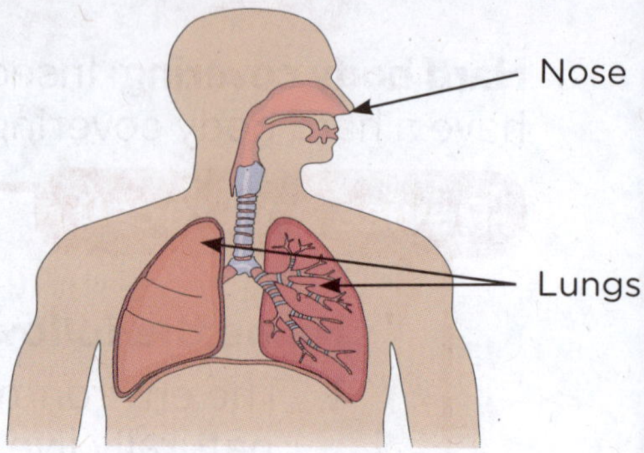

Breathing organs in humans

Lungs: Animals such as reptiles, birds and mammals take in air through their nose. The air that enters through the nose is carried to structures called **lungs**.

Science in Context

Mammals like dolphins and whales that live in water also breathe through lungs.

Amphibians such as frogs have a pair of lungs that help them to breathe on land and their moist skin helps them to breathe in water.

Frog

Organs for Movement

Animals move around in search of food and shelter, to look for their mate and to escape from their enemies.

Movement in Land Animals

Majority of the animals that live on land move with the help of legs (two at the front called **forelimbs**, and two at the back called **hindlimbs**). Kangaroos and frogs are good leapers. They have well-developed hindlimbs and short forelimbs.

Snails and slugs glide on a large single foot. They secrete[1] a sticky substance that helps them in gliding.

Word Web

[1]**Secrete**: to produce and release

Snail

Movement in Aquatic Animals

Aquatic animals such as fish move in water with the help of **fins** and **tail**.

Mammals such as dolphins, whales and porpoises have their forelimbs modified into **flippers**. Seals and penguins also swim with the help of these flippers.

The **webbed feet** of frogs help them to swim. Alligators and crocodiles use their legs as paddles and use their tail for swimming.

Squids, octopuses and cuttlefish move by **jet propulsion**. They take in water and expel it with great force. When the water is expelled backwards, these animals move forward.

Penguin

Seal

Movement in Insects

Insects are small animals with six legs. Most insects have wings that help them to fly, while some do not have wings (e.g. spiders, bedbugs and ants) and move with the help of legs.

Insects such as grasshopper hop with the help of their hind legs. Water boatman (a water insect) swims in water using its legs as oars[1].

Grasshopper

Water boatman

Word Web

[1]**Oar**: a long shaft of wood with a broad blade used for rowing of boats

Movement in Birds

Birds fly with the help of their wings (forelimbs) covered with feathers.

The wings also have strong muscles that assist birds in their flight. These muscles are called **flight muscles**.

Movement in birds

Science in Context

Bat is the only mammal that can fly. Its wings are not made of feathers but of membranous skin stretched and connected to the forelimbs.

Movement in Reptiles

Animals such as lizards and crocodiles crawl using their short legs.

Snakes do not have legs. They move with the help of scales and a flexible backbone, by pushing their body against the ground in a zig-zag manner.

Movement of a crocodile

Movement of a snake

Movement in Humans

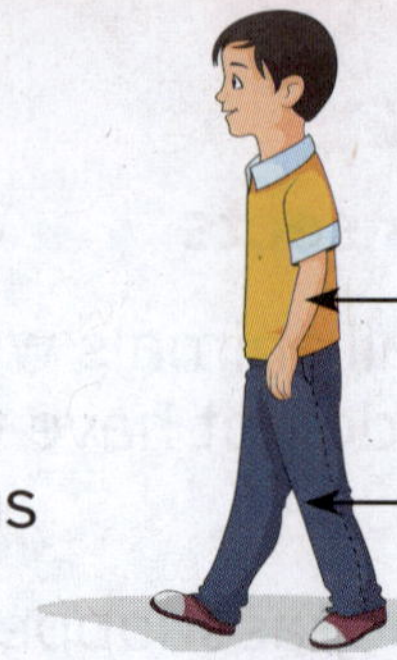

Movement of human beings

Human beings use their legs (hindlimbs) to stand, walk and run. The hands (forelimbs) and the arrangement of fingers help human beings to hold objects and do various activities.

MIGRATION

Migration of Siberian cranes

The movement of animals in large groups from one place to another in search of food and warmth is called **migration**.

Some birds migrate during the winter season to reproduce and find food. They use the fat stored in their bodies on such long journeys.

Siberian white cranes or snow cranes migrate to India during winters from Western Siberia.

Apart from birds, some mammals such as whales migrate during winter from the polar regions to warmer oceans.

Some insects such as monarch butterflies migrate during winter from Canada to Mexico.

Salmons (a type of fish) migrate from seas to very long distances to lay their eggs in rivers.

Science in Context
The Arctic tern is a seabird that travels between the Arctic and Antarctic regions, which is almost 35,000 km in distance, in a year. It is also called the 'champion of migration'.

Revise and recall

State whether the following statements are true (T) or false (F).

1. Prawns breathe with the help of gills. ☐
2. Insects have only two spiracles for breathing. ☐
3. Gills are the breathing organs of mammals. ☐
4. Seals swim with the help of flippers. ☐
5. A grasshopper hops with the help of its hind legs. ☐

Let's Say Together

Adaptation	Scales	Moulting
Spiracles	Flipper	Blubber
Chitin	Jet propulsion	Migration

Keywords

Habitat: The environment where an organism naturally lives and reproduces
Adaptation: The features and habits that help organisms to survive in their habitats
Moulting: The process of shedding old skin to replace it with new skin
Spiracles: The breathing organs of insects
Blubber: A thick layer of fat under the skin of animals that live in very cold places
Migration: The movement of animals in large groups from one place to another in search of food and warmth

Summary

1. Various habitats of animals include forest, desert, polar regions, mountain, ocean and freshwater.
2. Animals have various kinds of body coverings such as scales, shells, fur/wool and feathers to protect themselves from various climatic conditions.
3. Animals have different kinds of breathing organs such as skin, body surface, gills, spiracles and lungs.
4. Animals that live on land move with the help of four legs (two forelimbs and two hindlimbs).
5. Aquatic animals move with the help of fins, tail, flippers and webbed feet.
6. Feathers and wings help the birds in their movement.
7. Insects have legs and wings that help them to move.
8. Reptiles such as lizards and crocodiles crawl using their short legs.
9. The movement of animals in large groups from one place to another in search of food and warmth is called migration.

EXERCISE

A Choose the correct option for each of the following.

1. Identify the animal that belongs to the desert habitat.
   (a) Penguin (b) Seal (c) Camel (d) Walrus
2. Which of these is considered to be the largest habitat of the world?
   (a) Ocean (b) Forest (c) Desert (d) Freshwater
3. Jellyfish belong to which of these habitats?
   (a) Freshwater (b) Desert (c) Ocean (d) Polar
4. Identify the animal that has shell as its body covering.
   (a) Lizard (b) Tortoise (c) Yak (d) Crocodile
5. Which of these animals uses flippers for its movement?
   (a) Penguin (b) Lizard (c) Frog (d) Snake

B Fill in the blanks using the correct words.

1. The habitat present in Antarctica is __________________ . (polar/freshwater)
2. The body covering of fish is __________________ . (shell/scales)
3. The habitat with sandy lands and less rainfall is __________________ . (desert/forest)
4. Microscopic organisms breathe using their __________________ . (body surface/lungs)
5. ____________ use hindlimbs to stand, walk and run. (Humans/Crocodiles)

C Match the following.

Column A	Column B
(a) Frog	(1) Spiracles
(b) Dolphin	(2) Moist skin
(c) Cockroach	(3) Gills
(d) Fish	(4) Lungs and moist skin
(e) Earthworm	(5) Lungs

D **State whether the following statements are true (T) or false (F).**

1. The thick fur of polar bears keeps them warm in cold weather. ☐
2. Snakes do not have legs. ☐
3. All insects have wings. ☐
4. Water boatman swims in water using its legs. ☐
5. None of the birds migrate during the winter season. ☐

E **Give two examples for each of the following.**

1. Animals that live in oceans ________ ________
2. Animals that live in forests ________ ________
3. Animals that have shell as their body covering ________ ________
4. Animals that breathe through spiracles ________ ________
5. Reptiles that crawl using their short legs ________ ________

F **Answer the following questions briefly.**

1. What are the different types of habitats found on the Earth?
2. What is blubber?
3. Define adaptation.
4. What is moulting?
5. How do aquatic animals such as fish breathe in water?

G **Answer the following questions.**

1. Define habitat and write the main features of any two habitats.
2. Explain any three types of body coverings of animals with examples.
3. Describe any three types of breathing organs in animals with examples.
4. Write a note on the how aquatic animals and insects move.
5. Define migration. Why do animals migrate? Explain with examples.

H **HOTS (Higher Order Thinking Skills)**

1. How is the movement of penguin different from that of other birds?
(Hint: Organs for movement in both are different)
2. How is a whale different from a fish even though both are aquatic animals?
(Hint: Recall about the breathing organs in fish and whales.)
3. Why do birds migrate during winter? (Hint: Change in climatic conditions)

I **Look at the pictures given alongside. Circle the odd one based on their body coverings.**

J **Project**

1. Collect pictures of various animals from old newspapers and magazines. Group them according to their habitat and paste them in your scrapbook. Write a few lines about the adaptations of the animals in each habitat.
2. Make a poster on 'Migration'. Draw or paste the pictures of migratory animals and explain the reasons for migration. Display the poster in your classroom.

K **Cross-curricular Learning Link**

Locate the various ocean habitats: Pacific Ocean, Arctic Ocean, Atlantic Ocean, Southern Ocean and Indian Ocean in the world map given below. Colour them in blue.

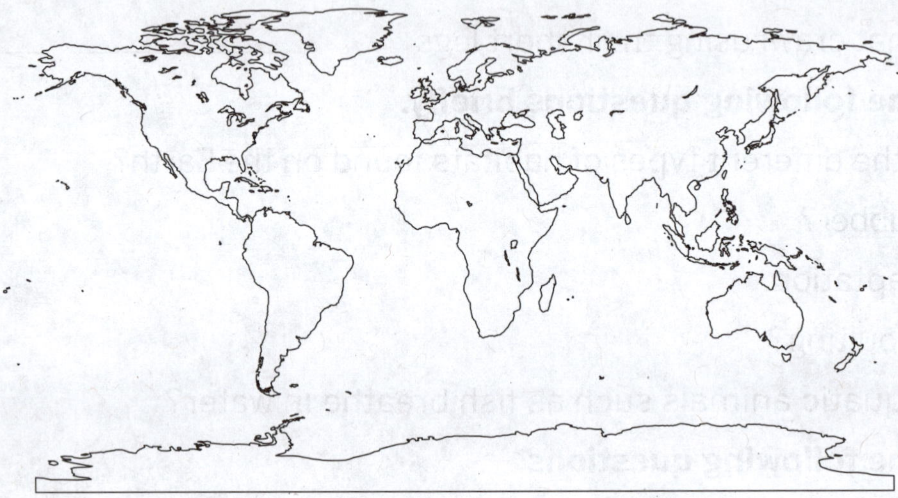

L **Life Skills**

Which of the following pictures show the correct way of treating animals/plants/habitat? Put a tick mark (✓) against the correct action and cross (✗) against the wrong action.

Planting a sapling ☐

Feeding a dog ☐

Boy throwing garbage in a pond ☐

Complete the table by filling in the habitat, body covering, breathing organ and type of movement in animals given below. One has been done for you.

Animal	Habitat	Body covering	Breathing organ	Movement
	Land/ desert	Scales	Lungs	Zig-zag movement

3 Skeletal System and Nervous System

Learning Checklist

- Skeletal system
- Nervous system
- Sense organs

Let's Get Started

Look at the pictures given below and fill in the blanks with the help of the words in the box.

brain	muscles	bones

I am able to run and stretch using the ____________ and ____________ in my body.

I am able to think and solve sums using my ____________ .

All the bones in our body are part of the skeletal system. We use our brain to think, read, write, etc. The brain is part of our nervous system. In this chapter, we will learn more about the skeletal system and the nervous system.

SKELETAL SYSTEM

The skeletal system consists of bones and cartilage. There are 206 bones in an adult human body. The framework of bones is called the **skeleton**, which gives support and shape to the body.

Following are some of the functions of the human skeletal system:

- It protects the soft internal organs of our body.
- It gives strength, support and shape to the body, and also helps in movement.
- It contains important minerals needed by our body.
- Many bones are hollow and filled with a substance called **bone marrow**. Bone marrow produces the blood cells.

The main parts of the skeletal system are the skull, the backbone, the ribcage and the limbs.

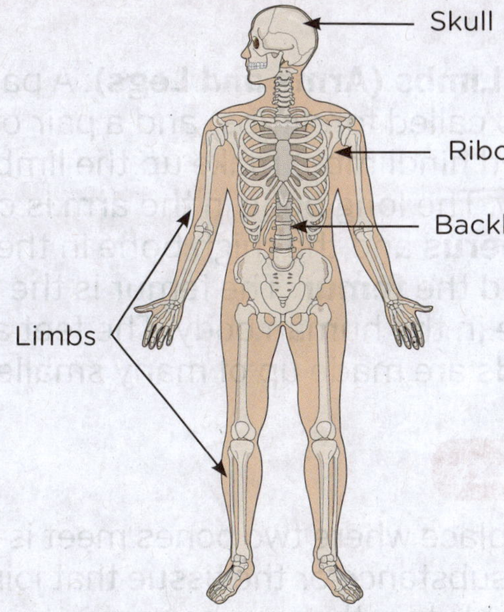

The human skeletal system

The Skull: The hard structure that you feel when you touch and press your head is called the skull. It protects the brain. The skull has 22 bones. It also has holes where the eyes, nose and ears are positioned. Of all the bones in the skull, only the lower jawbone can move; all others are fixed. The upper and lower jaws have teeth.

The Backbone (Spine): The long bone that you feel when you run your hand through the centre of your back is called the backbone, spine or the vertebral column. The backbone is made up of 33 small bones called the **vertebrae**. These bones help us in bending, twisting and turning. It also helps in keeping the body upright. The backbone protects the spinal cord inside it, which is connected to the brain in the skull.

The Ribcage: The frame of bones that forms a cage-like structure in the chest area is called the ribcage. It protects the heart and the lungs. It is made up of 12 pairs of long curved bones called the **ribs**. These ribs join the backbone at the back. Ten pairs of ribs are joined to the front bone called the breastbone or **sternum**. The remaining two pairs of ribs that are not attached to the breastbone are called the **floating ribs**.

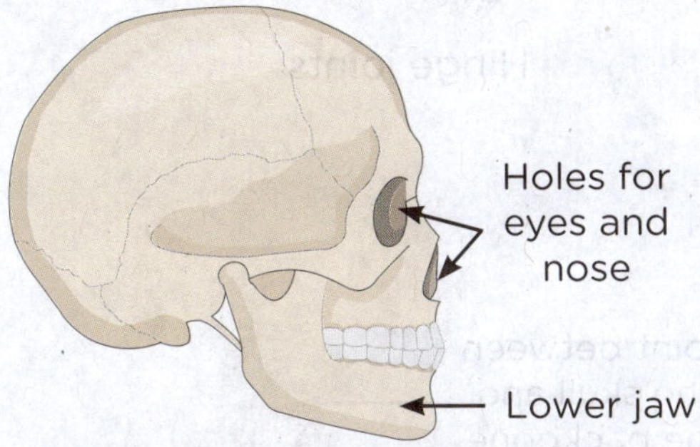

Skull

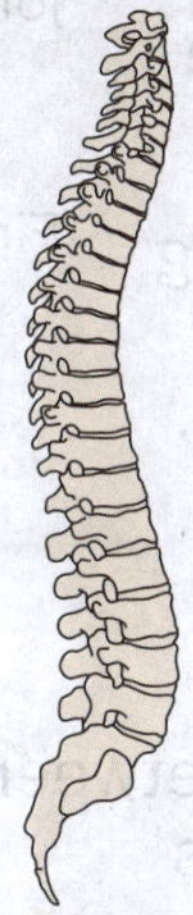
Backbone

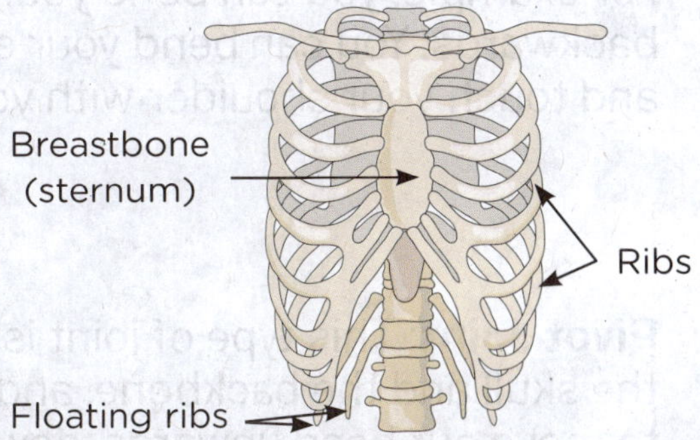

Ribcage

The Limbs (Arms and Legs): A pair of arms (also called forelimbs) and a pair of legs (also called hindlimbs) make up the limbs in our body. The long bone in the arm is called the **humerus** and the thigh bone in the leg is called the **femur**. The femur is the longest bone in the human body. The feet and the hands are made up of many smaller bones.

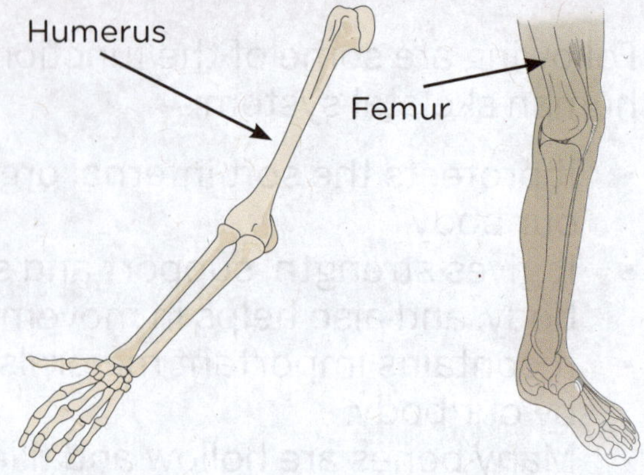

Limbs

Joints

The place where two bones meet is called a **joint**. The substance or the tissue that joins two bones or holds together two bones in place is called a **ligament**.

All the joints in our body are movable except the joints in the skull.

Joints are of different types and show different kinds of movements:

Ball and Socket Joint: This type of joint is found in the **shoulder** and **hip** regions, and allows movement of these parts in all directions.

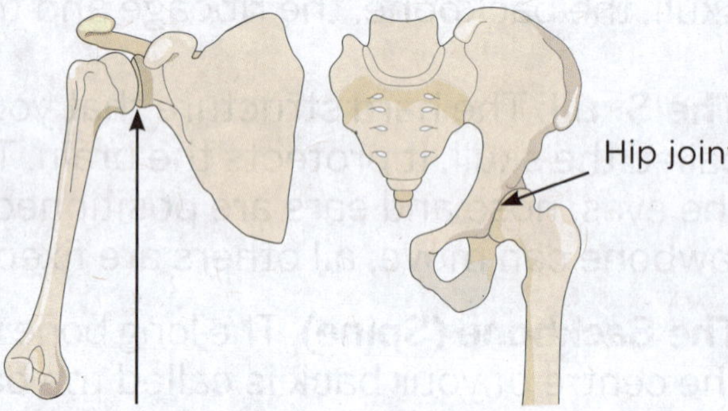

Ball and socket joints

This joint helps in movements such as bending, stretching and rotating.

Hinge Joint: This type of joint is found in the elbows, fingers, knees and toes, and helps us to move these parts in only one direction.

For example, you can bend your lower leg backwards. You can bend your elbows and touch your shoulder with your hand.

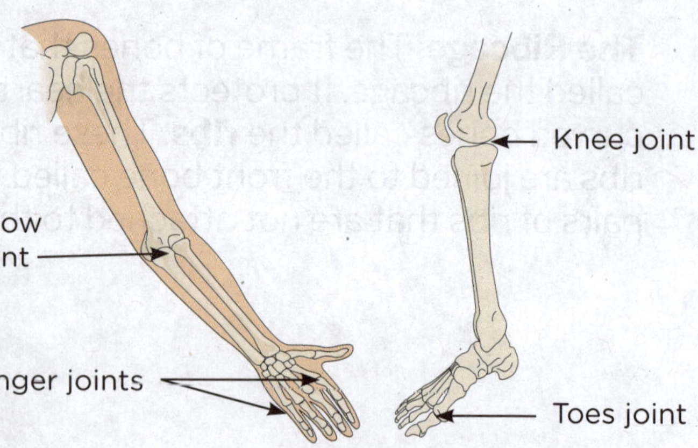

Hinge joints

Pivot Joint: This type of joint is found between the skull and the backbone, and helps us to move our head upwards, downwards and sideways.

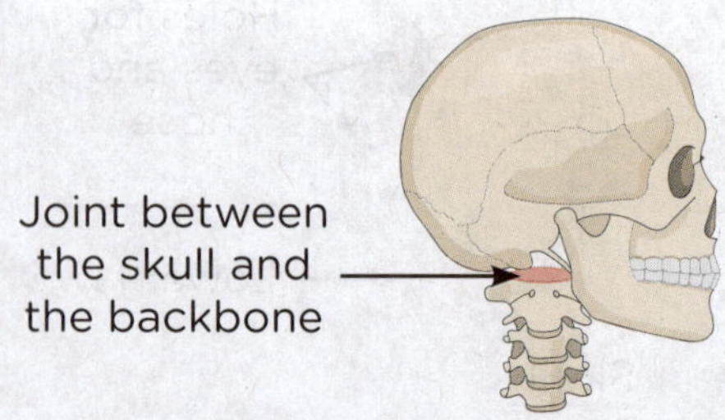

Pivot joint

Gliding Joint: This type of joint is present in the ankles and the wrists, and helps the bones of these parts to slide easily.

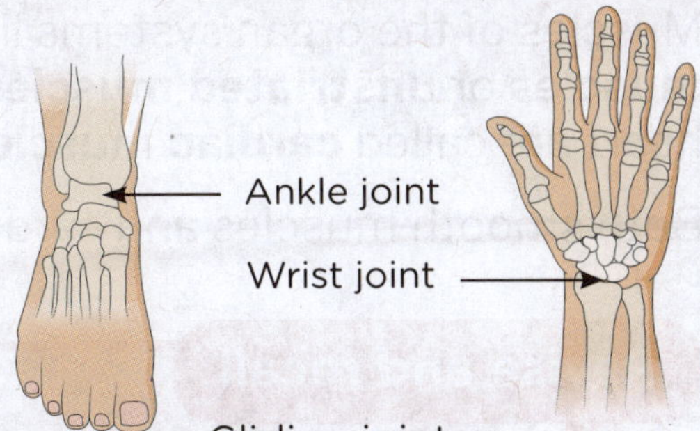

Gliding joints

Cartilage

Cartilage is a tough, elastic tissue that is found in various parts of the body such as the joints in the ankle, knee and elbow and also between the vertebrae in the backbone. Cartilage prevents the bones from rubbing against each other and wearing out.

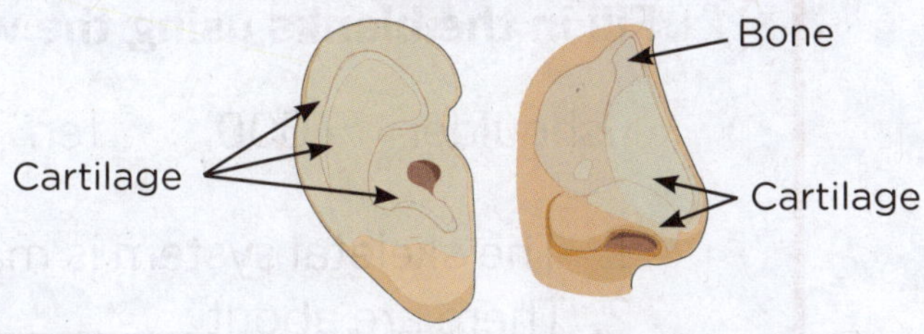

Cartilage in the ear and the nose

Cartilage is also found in the ears and the nose. Touch the top part of your ear or the tip of your nose. These parts are soft to touch. They are the cartilages.

Science in Context

A baby's skeletal system is mostly made up of cartilage. As the baby grows, most of the cartilage is replaced by bones.

Bones and cartilage alone cannot bring about movement. Movement happens when bones and cartilage work together with muscles.

Let us learn how bones and muscles work together.

Muscles

There are more than 600 muscles in our body. The bones in our body cannot move without the help of muscles. Muscles are attached to the bones with the help of tough cords called **tendons**. To move a bone, two muscles come into play. When you want to move your arm or leg, one muscle has to contract and another muscle has to relax to allow movement of the arm or leg.

Muscles are of two types: voluntary and involuntary.

Voluntary Muscles: The muscles that are under our control are called voluntary muscles.

All the muscles attached to the skeleton (called **skeletal muscles**) of the body are voluntary muscles, for example, muscles in the arms and legs.

When looked through a microscope, these muscles seem to have bands or stripes on them. Therefore, they are also called **striped** or **striated muscles**.

Involuntary Muscles: The muscles that are not under our control are called involuntary muscles.

Muscles of the organ systems like the digestive system are called **smooth muscles** or **unstriated muscles** because they are not striped. Muscles of the heart are called **cardiac muscles**.

Both smooth muscles and cardiac muscles are examples of involuntary muscles.

Revise and recall

Fill in the blanks using the words from the help box.

shoulder 600 ten 206 33 tendon skeleton

1. The skeletal system is made up of __________ bones.
2. There are about __________ muscles in our body.
3. __________ pairs of ribs are joined to the breastbone.
4. The backbone is made up of __________ small bones.
5. __________ is a tough cord that connects the muscle to the bone.
6. The joint in the __________ allows movement in all directions.
7. Voluntary muscles are attached to the __________ in the body.

NERVOUS SYSTEM

What helps you to remember the topics that you have read in your previous chapters or classes?

How do you blink your eyes without even knowing about it?

We are able to do these tasks with the help of the nervous system.

The nervous system is made up of the brain, the spinal cord and a network of nerves that run throughout our body.

The nervous system controls all the other organ systems of our body.

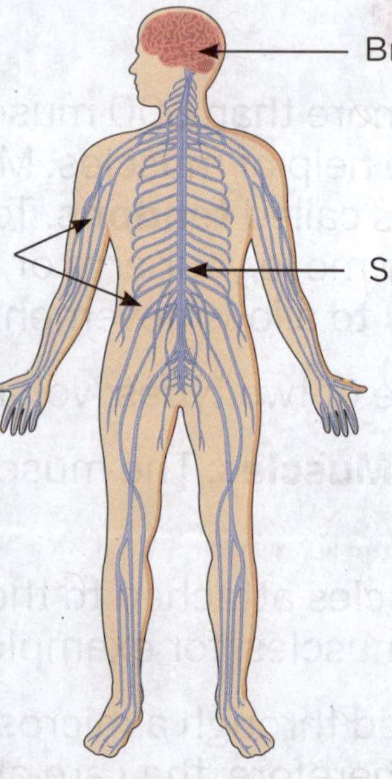

The nervous system

Brain

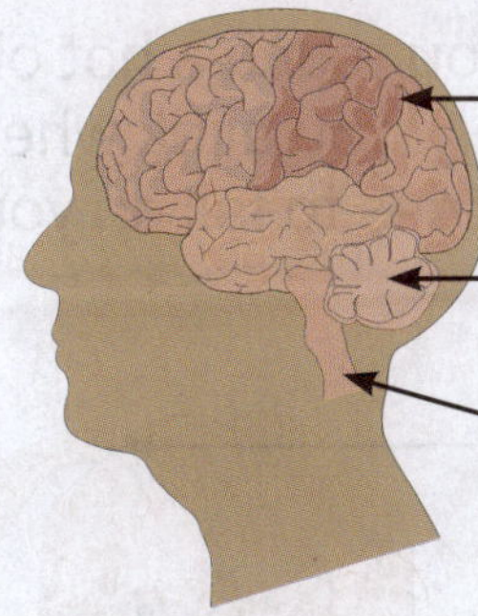

The human brain

The brain controls everything in the human body. The skull protects the brain. The part of the skull that protects the brain is called the **cranium**.

The brain is made up of three parts that carry out different tasks. They are the cerebrum, the cerebellum and the brain stem.

Cerebrum: This is the largest part of the brain. It controls our intelligence, memory, logic and sense organs. It also controls our body movements.

Cerebellum: This is the second largest part of the brain and is present behind the cerebrum. It controls balance and posture of our body, and also actions of the muscles in the body.

Brain Stem: The brain stem connects the brain to the spinal cord. The nerves in the brain stem control our heartbeat, breathing and the blood pressure.

Science in Context
The average adult human brain weighs about 1.3 to 1.4 kg.

Spinal Cord

The spinal cord is a bunch of nerves that begins from the base of the brain and connects it to other parts of the body.

It is protected inside the backbone. Its main function is to pass on information between the brain and the other body parts.

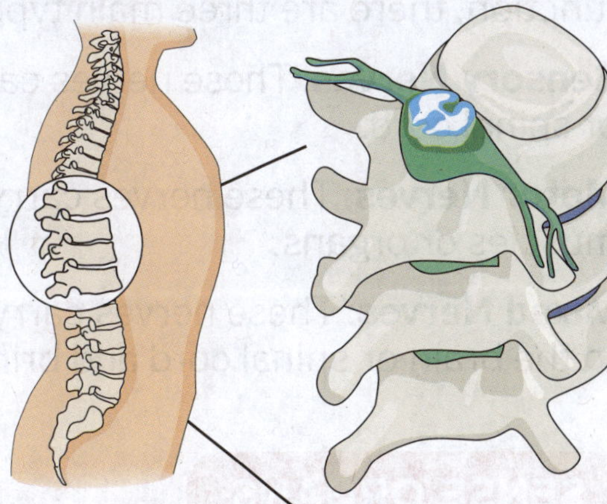
Spinal cord inside the backbone

Reflex Actions

A reflex action is an involuntary or automatic action of our body in response to something.

Most reflex actions take place when the messages are sent to and from the spinal cord, without involving the brain. Hence, such responses are very fast.

For example, if you touch a hot object by mistake, a reflex causes you to remove your hand immediately. You blink when something flies in front of your eyes; you step back when someone tries to fall on you. Even coughing and sneezing are reflexes.

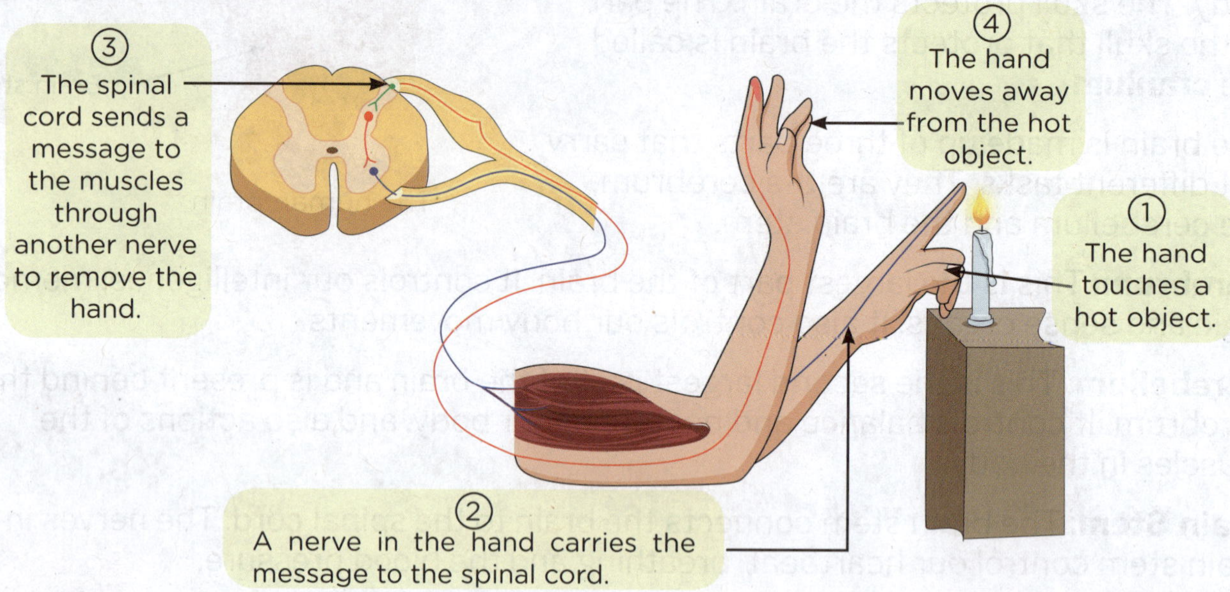

Reflex action

Nerves

Nerves are bundles of long, thin, thread-like structures that carry messages between the brain or spinal cord and other parts of the body. Depending on their function, there are three main types of nerves in our body.

Sensory Nerves: These nerves carry messages from the sense organs to the brain or spinal cord.

Motor Nerves: These nerves carry messages from the brain or spinal cord to the muscles or organs.

Mixed Nerves: These nerves carry out both the functions, that is, carry messages to the brain or spinal cord and bring back messages from the brain or spinal cord.

SENSE ORGANS

Sense organs are the organs that give messages about our surroundings to our nervous system. The sense organs include the eyes, the ears, the nose, the tongue and the skin.

The Eyes: We have two eyes to help us see. The coloured part of our eye is called the **iris**. The black circle in the centre of the eye is called the **pupil**, which allows light to enter through it.

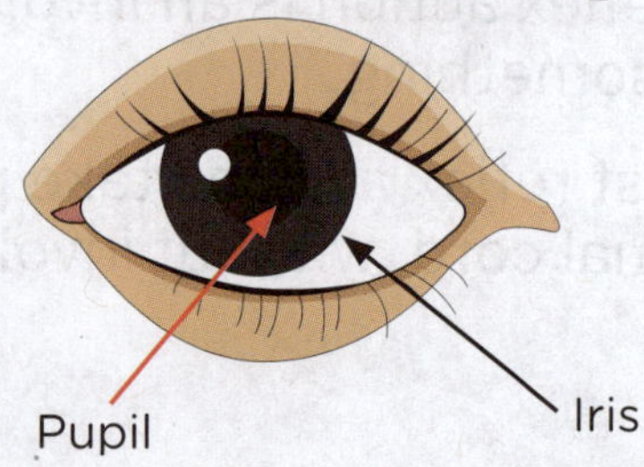

A transparent, dome-shaped part called the **cornea** covers the front portion of the eye and protects the pupil and the iris. The lens is present behind the iris.

The inner part of our eye has a very thin layer of tissues called the **retina** which captures the light that enters the eye and sends it to the brain through the **optic nerve**.

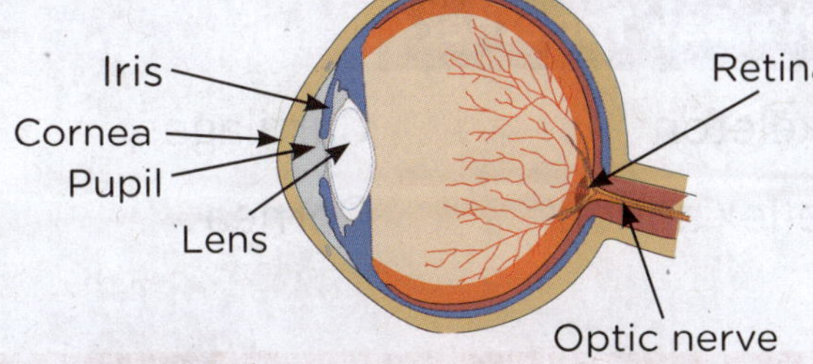

The human eye

The Nose: The nose helps us to smell and breathe. The nose has two holes called **nostrils**.

At the back of the nose, there is a space called the **nasal cavity**. It is separated from the mouth by a bony structure called the **palate**.

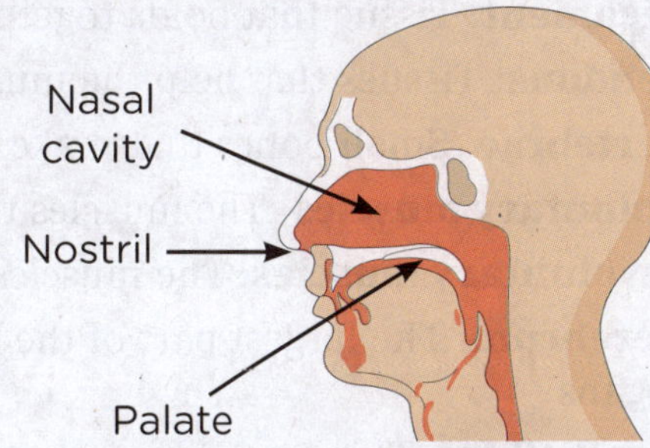

The nose

The Ears: The ears help us to hear sounds. The outer part of the ear that we can see is called the **pinna**. It collects sound and sends it to the **eardrum**. The eardrum sends vibrations to the **inner ear** through the **middle ear**.

The inner ear converts these vibrations to a signal which is sent to the brain.

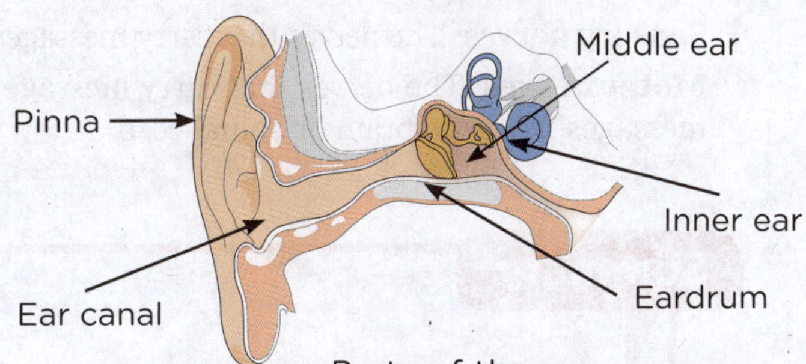

Parts of the ear

The Tongue: The tongue is a fleshy sense organ found on the floor of the mouth.It helps us to speak, chew and taste.

The taste buds on the tongue help us to identify different tastes by sending signals to the brain.

The primary tastes that the tongue can sense are sweet, salty, sour and bitter.

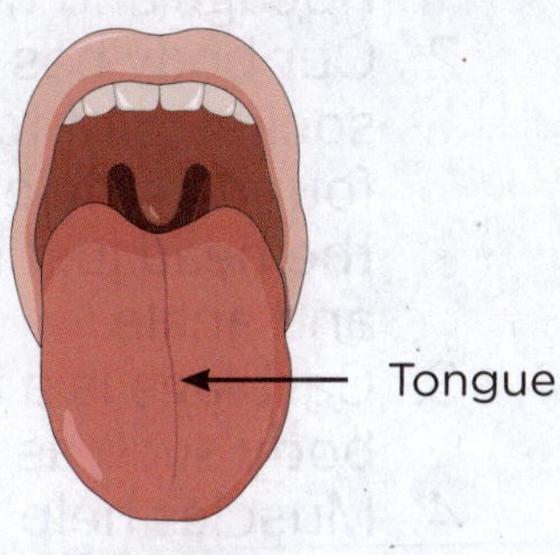

The Skin: The skin, which is the largest organ and the outer covering of the body, can sense touch and helps us to feel heat, cold, pain, etc.

The skin carries out many functions needed for our survival which include wound healing, protection, maintaining the body temperature and preventing infections.

Science in Context

Butterflies can taste with their feet as they have taste buds situated in their feet.

Let's Say Together

Skeleton	Cartilage	Cerebrum	Cerebellum
Reflex action	Cornea	Retina	Pinna

Keywords

Ligament: Tissue that holds together two bones in place

Tendons: Tissues that help the muscles to attach to the bones

Vertebrae: Small bones that make up the backbone

Voluntary muscles: The muscles that are under our control such as muscles in the arms and legs

Involuntary muscles: The muscles that are not under our control such as the heart muscles

Cerebrum: The largest part of the brain that controls our intelligence, memory, logic and sense organs

Cerebellum: The second largest part of the brain that controls balance and posture of our body, and also actions of the muscles

Reflex action: The involuntary or automatic action of our body in response to something

Sensory nerves: The nerves that carry messages from the sense organs to the brain or spinal cord

Motor nerves: The nerves that carry messages to the brain or spinal cord and also bring back messages from the brain or spinal cord

Summary

1. The skeletal system consists of the skull, the backbone, the ribcage and the limbs.
2. Our body has four different types of movable joints: ball and socket joint found in the hip and shoulder regions; hinge joint found in the elbow, fingers, knee and toes; pivot joint found behind the head just below the skull; and gliding joint found in the wrist and ankle.
3. Cartilage is a tough, elastic tissue found in various parts of the body such as the joints, the ears and the nose.
4. Muscles help to move the bones in our body and are attached to the bones by tough cords called tendons.
5. The nervous system controls all the other organ systems of our body using the brain, the spinal cord and a bundle of nerves.
6. The spinal cord controls the automatic actions of the body which are called reflex actions.
7. Sense organs carry messages about our surroundings to our nervous system. The sense organs include the eyes, the ears, the nose, the tongue and the skin.

EXERCISE

A Choose the correct option for each of the following.

1. The function of the skeletal system is to give __________ to the body.
   (a) support (b) shape (c) strength (d) All of these
2. The joint between the skull and the backbone is __________.
   (a) ball and socket (b) pivot (c) hinge (d) gliding
3. Ten pairs of ribs are connected in the front to the __________.
   (a) backbone (b) sternum (c) brain (d) nerves
4. Which of these structures protects the brain?
   (a) Backbone (b) Ribcage (c) Skull (d) Forelimb
5. Most reflex actions that take place in our body are controlled by the __________.
   (a) brain (b) spinal cord (c) ribs (d) joints

B Fill in the blanks using the correct words.

1. __________ protects the soft internal organs of our body. (Nerves/Skeleton)
2. The __________ controls everything that the body does. (brain/bones)
3. __________ muscles are attached to the skeleton. (Voluntary/Involuntary)
4. Most of the joints in the skull are __________. (movable/immovable)
5. The nasal cavity is separated from the mouth by a bony structure called __________. (cranium/palate)

C Match the following.

Column A	Column B
(a) Humerus	(1) Hinge joint
(b) Ankle	(2) Tissue that joins two bones
(c) Hips	(3) Gliding joint
(d) Knees	(4) Ball and socket joint
(e) Ligament	(5) Long bone in the arm

D State whether the following statements are true (T) or false (F).

1. Bones and muscles work together to bring about movement. ☐
2. Gliding joints are found between the small bones in the vertebral column. ☐
3. The nervous system is made up of the brain, the heart and the nerves. ☐
4. The brain stem controls our intelligence, memory, logic and sense organs. ☐
5. The skin is the largest organ in our body. ☐

E Name the following.

1. The longest bone in the human body ______________
2. A place where two bones meet ______________
3. The type of joint in the shoulder ______________
4. The system that controls everything you do ______________
5. The part of the brain that helps us to maintain balance of our body ______________

F Answer the following questions briefly.

1. How many pairs of ribs are there in a human body? What are floating ribs?
2. What are joints? Name the different types of joints found in a human body.
3. What is cartilage? Where is it found in the body?
4. Name the three parts of the brain. Which is the largest part of the brain?
5. What is a reflex action?
6. Name the different types of nerves.

G Answer the following questions.

1. What are the functions of the human skeletal system?
2. Name the main parts of the skeletal system. Explain the structure of the ribcage and the spinal cord.
3. Describe the two types of muscles in our body. Give examples for each.
4. What are the functions of the different parts of the brain?
5. Explain how a reflex action takes place when you touch a hot pan.
6. What are sense organs? Explain briefly the functions of each sense organ.

H **HOTS (Higher Order Thinking Skills)**

1. What would happen if we do not have any bones in our body?
(Hint: Think about the functions of the skeletal system.)
2. Why do people find it difficult to move their body as they get older?
(Hint: Bones grow weaker with age.)

I **Label the different parts of the brain in the following diagram.**

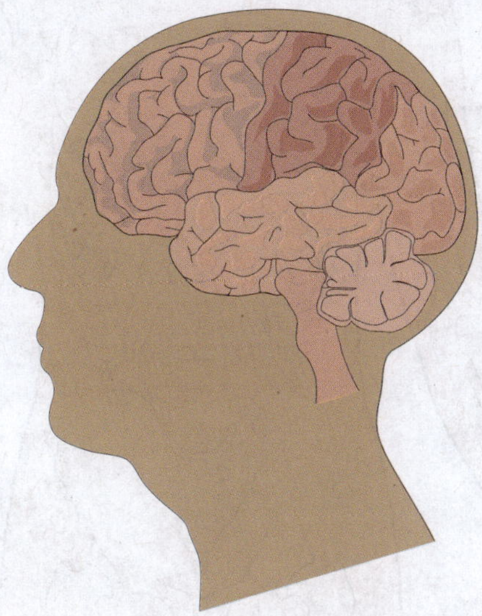

J **Project**

Prepare a chart on the different types of joints present in our body and display the chart in your class. Also, write the kind of movement shown by each joint.

K **Life Skills**

A diet that has food grains, vegetables, fruits, protein and small amounts of milk and fats helps to keep our skeletal and nervous systems healthy.

Exercise and good sleep keep the brain active and the body fit. Exercises such as brisk walking, swimming, running, playing and dancing help to maintain a healthy body weight and make the bones stronger.

Write five healthy habits in your notebook that you will follow to keep your skeletal and nervous systems healthy.

L **Cross-curricular Learning Link**

Write a story in 100 words about a dish you made in the kitchen. Use the terms related to different tastes that the tongue can sense in your story.

WORKSHEET

Label the parts of the skeletal system and the types of joints that you have studied in this chapter.

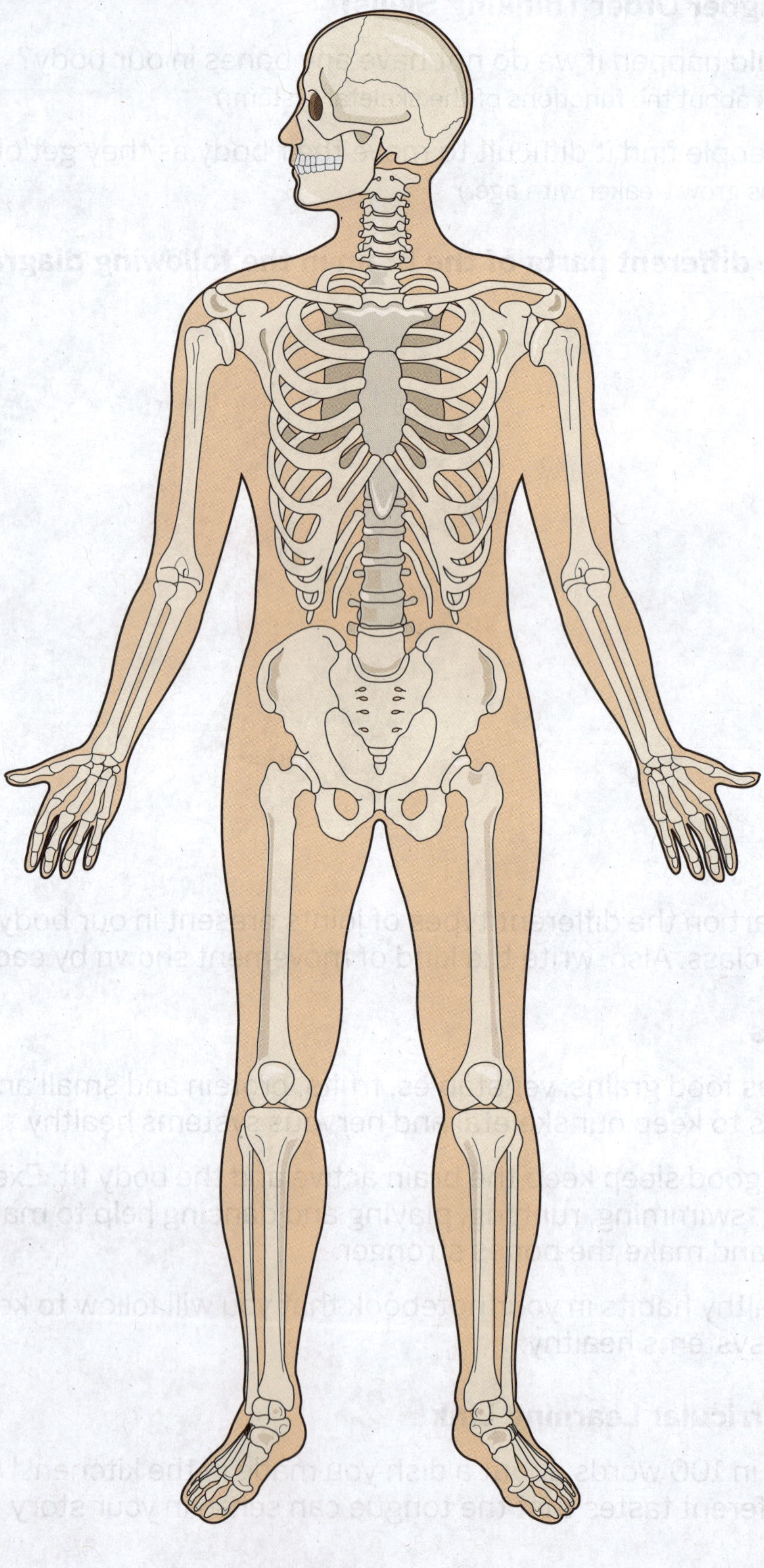

4 Food and Health

Learning Checklist

- Components of food
- Balanced diet
- Diseases

Let's Get Started

We get food from both plants and animals. Look at the pictures of the food items given below and name the sources (plant or animal) from which they are obtained.

Eggs

____

Rice

____

Potato fries

____

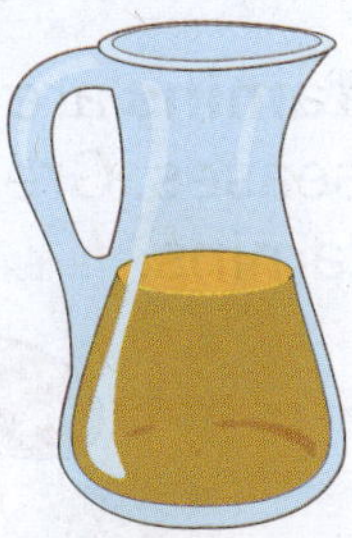
Orange juice

____

Cheese

____

Bread

____

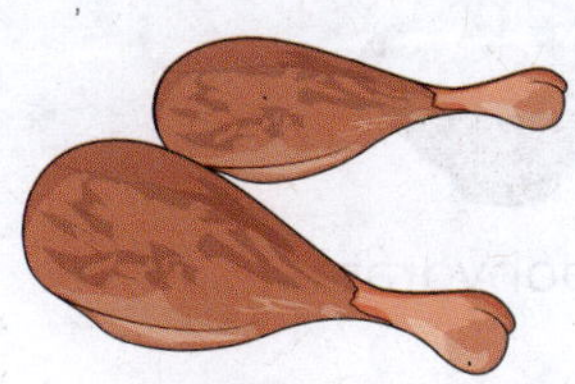
Chicken

____

Ice cream

____

Have you ever thought why you eat different varieties of food and not just your favourite food item all the time?

In this chapter, we will learn about the need to eat a variety of food, and the importance of rest and exercise in keeping us healthy and fit.

COMPONENTS OF FOOD

Food contains nutrients[1] that are required by the body in adequate[2] amounts to grow and stay healthy. The main nutrients present in food are carbohydrates, fats, proteins, vitamins and minerals. Food also contains water and roughage.

Word Web

[1]**Nutrients**: components found in food that give strength and help the body to grow

[2]**Adequate**: sufficient in quantity

Carbohydrates, proteins and fats are called **macronutrients** as these are required in large quantities in our daily diet, while vitamins and minerals are called **micronutrients** as these are required in small quantities in our daily diet.

Carbohydrates are the main source of energy for our body. Rice, wheat, ripe fruits, bread, honey, sugar and potato are some food items rich in carbohydrates.

Fats also provide energy to the body. Butter, oil, ghee and nuts are rich sources of fats.

Proteins help in the growth and repair of tissues. They also help in building muscles. Groundnut, beans, pulses, fish, eggs, milk, cheese and curd are some food items rich in proteins.

Vitamins are required in small amounts to keep our body fit and protect it against diseases. Green leafy vegetables, carrot and fruits are some food items rich in vitamins.

Sources of carbohydrates

Sources of fats

Sources of proteins

Sources of vitamins

Sources of minerals

Minerals help in the physical and mental development of the body, and keep our bones and teeth strong and healthy. Milk, cheese, eggs, fruits and vegetables are some food items rich in minerals.

Roughage (also called dietary fibre) is the undigested portion of the plant food. It adds bulk to our diet. It helps in easy passage of food through the digestive system. Cereals, unpolished rice and grains, fresh fruits and vegetables are some rich sources of roughage.

Water is required for almost all functions of the body. We should drink plenty of water (at least 8–10 glasses daily) to stay healthy.

BALANCED DIET

A diet that has all the components of food in adequate amounts is called a **balanced diet**.

A balanced diet is essential for the proper growth and functioning of the body.

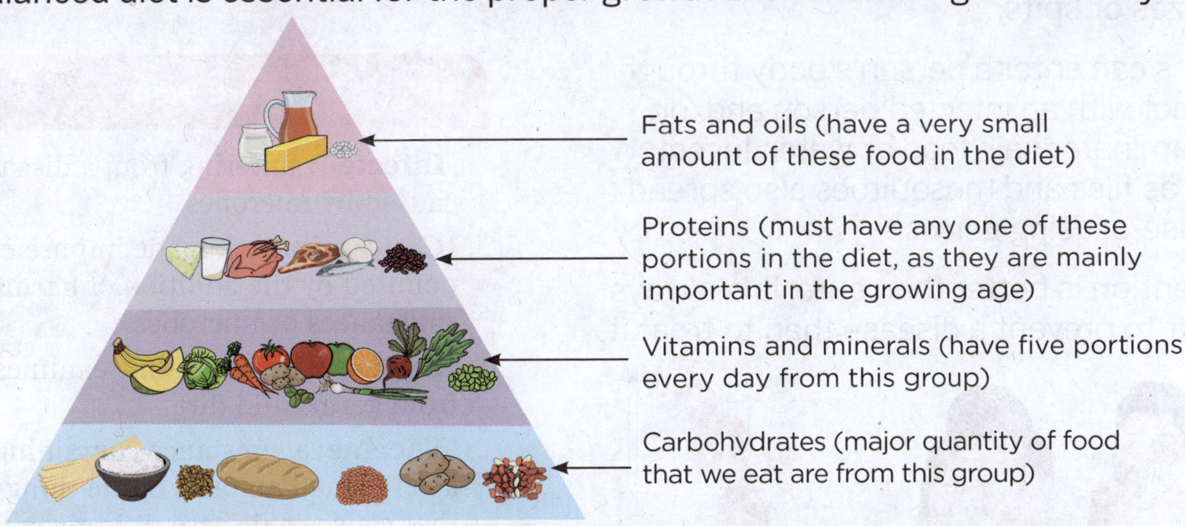

In addition to a balanced diet, our body needs regular exercise, adequate rest and sleep to stay healthy and active. Exercise keeps our bones and muscles strong. Rest and sleep keep us relaxed and refreshed.

Revise and recall

Fill in the blanks using the correct words.

1. Milk and milk products are rich in ______________ . (protein/fat)
2. ______________ protect our body from diseases. (Vitamins/Minerals)
3. Roughage is the ______________ portion of the plant food. (digested/undigested)
4. We should eat ________ portions of fruits and vegetables every day. (two/five)

DISEASES

Health is defined as a state of complete physical and mental wellbeing. A **disease** is a state where the body and/or mind are not in a healthy condition. A disease can occur due to many reasons. We can broadly classify diseases into two categories: communicable diseases and non-communicable diseases.

Communicable Disease

A disease that can be passed on from an infected[1] person to a healthy person is called **communicable disease**. Such diseases are caused by microbes such as bacteria, viruses, fungi and protozoa. Microbes that cause diseases are called **germs**. These germs can spread when the infected person coughs, sneezes or spits.

Coughing Sneezing Spitting

Germs can enter a person's body through contact with an infected person and/or contaminated[2] air, food or water. Insects such as flies and mosquitoes also spread disease-causing germs.

Prevention is better than cure. It is always better to prevent a disease than to treat it.

Cover your mouth while coughing and sneezing

Wash your hands before eating

Word Web

[1] **Infected**: suffering from a disease caused by microbes

[2] **Contaminated**: made impure or polluted by the addition of harmful substances or microbes

[3] **Hygiene**: to maintain cleanliness from germs and dirt

[4] **Vaccine**: a substance containing a very small amount of the thing that causes a disease. It is given to people to prevent them from getting that disease

Maintaining good hygiene[3] and keeping the surroundings clean play a very important role in preventing diseases.

Vaccination is the process of using substances called vaccines[4] for the prevention of diseases. Many communicable diseases can be prevented by vaccinations given at the right time. Vaccination protects a person by not allowing the germs to multiply in case they enter the body.

Cover food and water

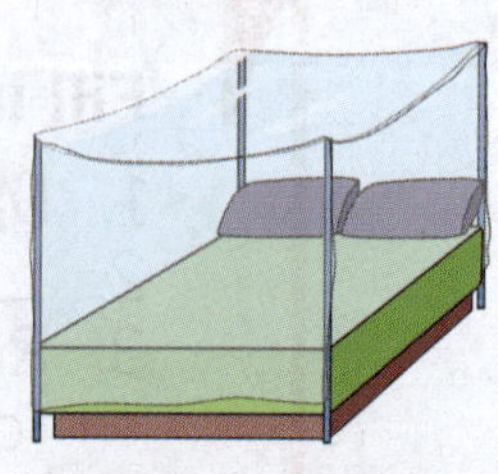

Use mosquito nets

Some communicable diseases, their symptoms[1] and how they can be prevented or treated are given below.

Word Web

[1]**Symptom**: sign of a particular disease

Communicable diseases	Caused by	Symptoms	Prevention/Cure
Flu	Virus	Sore throat, runny nose, difficulty in swallowing, fever, headache, cough and body ache	Take medicines, plenty of rest, plenty of fluids, cough drops, salt water gargling
Measles	Virus	Runny nose, fever, cough, redness in the eyes, rashes, weight loss	Vaccination at an early age, plenty of fluids and bed rest can decrease the symptoms to some extent
Chickenpox	Virus	Fever, headache, sore throat, tiredness, rashes that slowly dry up and then form scars	Vaccination at an early age, medicines, cool baths and bed rest. The rashes should not be scratched as they may spread to other places on the body.
Cholera	Bacteria	Vomiting, dehydration, diarrhoea	Drink lots of fluids, take medicines
Malaria	Protozoa	Fever, chills, sweats, headaches, vomiting and body ache	Medicines to be taken on time, drips, use mosquito nets or mosquito-repellent cream, avoid storing water in open containers
Typhoid	Bacteria	Headache, body ache, fever, diarrhoea	Take medicines and drips, boiled water, food without oil and spices, bed rest

Science in Context

The fresher the food we eat, the healthier we will be. Breakfast is the most important meal of the day as it gives our body and mind the required nutrients and energy to carry out all the functions for the rest of the day. We should never skip breakfast.

Non-communicable Diseases

A disease that cannot be passed on from one person to another person is called **non-communicable disease**.

Such diseases are mostly caused by the deficiency[1] of adequate amounts of nutrients in the diet. Hence, they are also called **deficiency disorders**.

Word Web

[1] **Deficiency**: lack of something

The following table gives some common deficiency disorders, their symptoms and the food that should be taken to prevent these diseases.

Deficiency disorders	Symptoms	Food sources of that nutrient
Anaemia (iron deficiency)	Weakness, fatigue, difficulty in breathing	Raisins, apricots, milk, meat, apples, green leafy vegetables
Night blindness (vitamin A deficiency)	Inability to see in poor light or at night	Carrots, sweet potato, animal liver, red capsicum, lettuce
Beriberi (vitamin B1 deficiency)	Weight loss, weakness, low heartbeat, tiredness, affects growth of body and muscle formation	Beans, cereals, peas, fish, nuts, watermelon
Scurvy (vitamin C deficiency)	Irritation, swelling of joints, bleeding gums	Guavas, papaya, citrus fruits, tomatoes, grapes, berries
Rickets (vitamin D deficiency)	Softening and weakening of bones	Cod-liver oil, fish, milk and dairy products, eggs, mushrooms, spinach
Goitre (iodine deficiency)	Deafness, tiredness, weight gain, swelling in the neck	Iodized salt, milk, boiled eggs, strawberries, yoghurt, baked potatoes
Kwashiorkor (protein deficiency)	Bloated belly, slow growth, weakness, weak development of limbs	Fish, eggs, meat, milk and other dairy products

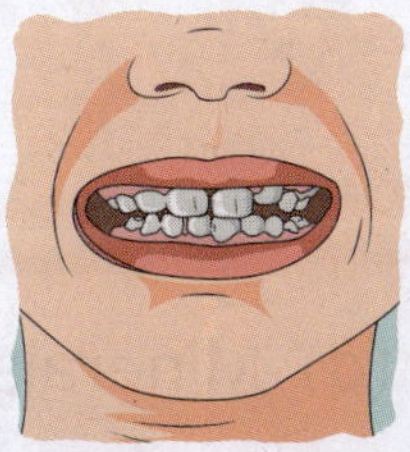

Vitamin C deficiency (scurvy)

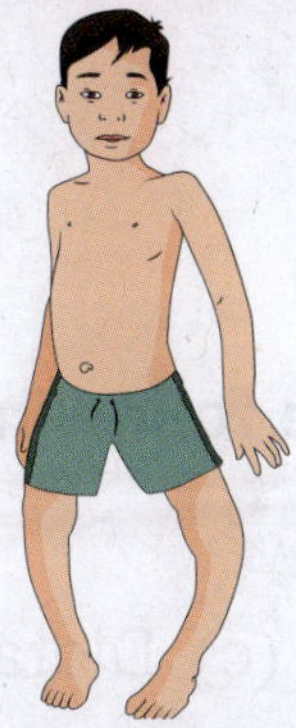

Vitamin D deficiency (rickets)

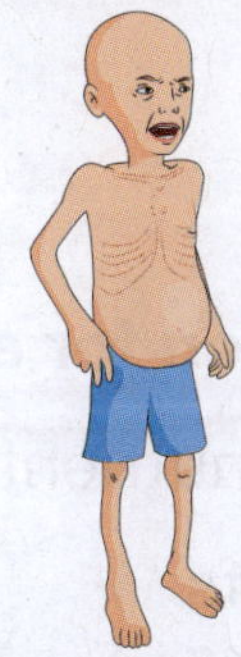

Protein deficiency (kwashiorkor)

Iodine deficiency (goitre)

Let's Say Together

Hygiene	Vaccination	Communicable	Non-communicable
Anaemia	Scurvy	Rickets	Kwashiorkor

Keywords

Macronutrients: Nutrients required in large quantities in our daily diet

Micronutrients: Nutrients required in small quantities in our daily diet

Roughage: The undigested portion of the plant food that adds bulk to our diet

Communicable disease: A disease that can be passed on from an infected person to a healthy person

Non-communicable disease: A disease that is not passed on from one person to another person

Vaccination: The process of using vaccines for the prevention of diseases

Summary

1. The five components (nutrients) present in our food are carbohydrates, proteins, fats, vitamins and minerals.
2. We should have a balanced diet by eating the adequate amounts of various foods along with water and roughage in order to grow and stay healthy.
3. Diseases can be broadly classified into communicable diseases and non-communicable diseases.
4. Many of the communicable diseases can be prevented at a very early age by vaccinations, while non-communicable diseases can be prevented by having a healthy, balanced diet.

EXERCISE

A Choose the correct option for each of the following.

1. Which of these are macronutrients?
   (a) Vitamins (b) Fats (c) Dietary fibres (d) Minerals
2. Our bones and teeth will be well developed when we eat foods rich in
   (a) fats (b) sugars (c) minerals (d) All of these
3. Which of the following is a communicable disease?
   (a) Typhoid (b) Malaria (c) Cholera (d) All of these
4. Vitamin B1 deficiency causes
   (a) weight loss (b) softening of bones (c) blindness (d) weight gain
5. Citrus fruits and tomatoes should be eaten to avoid
   (a) anaemia (b) night blindness (c) scurvy (d) goitre

B Fill in the blanks using the correct words.

1. Nutrients required in small quantities in our daily diet are called ____________. (micronutrients/macronutrients)
2. ____________ are required for the physical and mental development of the body. (Vitamins/Minerals)
3. We should drink at least ____________ glasses of water daily. (1–3/8–10)
4. Non-communicable diseases can be avoided by following a good ____________. (diet/hygiene)
5. Vomiting and dehydration are the symptoms of ____________. (cholera/chickenpox)

C Match the following.

Column A	Column B
(a) Iodine deficiency	(1) Scurvy
(b) Vitamin C deficiency	(2) Goitre
(c) Vitamin B1 deficiency	(3) Anaemia
(d) Iron deficiency	(4) Rickets
(e) Vitamin D deficiency	(5) Beriberi

D **State whether the following statements are true (T) or false (F).**

1. Proteins are the building materials of our body parts. ☐
2. Vaccination can prevent communicable diseases. ☐
3. Viruses and fungi cause non-communicable diseases. ☐
4. We can have a balanced diet when we eat a variety of food in adequate amounts. ☐
5. Plenty of fluids and bed rest can decrease the symptoms of measles to some extent. ☐

E **Give two examples for each of the following.**

1. Sources of carbohydrates ________ ________
2. Sources of fats ________ ________
3. Sources of vitamins ________ ________
4. Sources of minerals ________ ________
5. Sources of roughage ________ ________

F **Answer the following questions briefly.**

1. What is roughage?
2. Give any two habits that should be followed to prevent communicable diseases.
3. Name two communicable and non-communicable diseases.
4. Why is vaccination given?
5. What are the symptoms of night blindness?

G **Answer the following questions.**

1. What are the different components of food? Also mention their important sources.
2. What are micronutrients and macronutrients? On what basis are they grouped?
3. What is a balanced diet? Why are water and roughage important in a balanced diet even when they do not contain any nutrients?
4. Make a chart of the nutrients required for a balanced diet and in what quantity should they be eaten.
5. What are communicable diseases? What are the symptoms of malaria and how can it be prevented?
6. What are deficiency disorders? What are the symptoms of rickets and how can it be prevented?

H HOTS (Higher Order Thinking Skills)

1. How can a person with a cold or flu avoid spreading germs?
(Hint: Good hygiene)
2. If you had been to a party and eaten a lot of junk food and sweets, how would you compensate for it in your other meals for that day?
(Hint: Balanced diet)
3. Manav mostly has a good diet but still feels lazy and gets tired easily. He does not exercise or play any outdoor games. He spends his weekends watching television or playing computer games. He often goes to bed late at night. What should Manav do to be active and fit?
(Hint: Apart from a balanced diet, what else is required to stay healthy and active?)

I Identify the nutrients present in the food items given below.

________________ ________________ ________________

J Project

1. Make a food diary. List all the five nutrients in your diet that you will eat on one particular day. Jot down the portions of each food item you would have eaten for that day and present it in your class.

   Ask your teacher to check your food diary and advise if you are following the right diet.
2. **Pick and speak**: Distribute chits in the class with the names of food items listed below. Each student can pick up a chit and speak on the nutrients present in that food item.

Butter, papaya, sweet lime, paneer, watermelon, fish, carrots, boiled eggs, green leafy vegetables, cereals, rice

K Life Skills

Having pets in our homes brings us joy and teaches us values of showing kindness to them. But we also have to keep in mind that not only pets but also insects and small reptiles that live in and around our homes may carry diseases that can spread when we touch them or through the air, water or objects that they touch or sit on.

From the list given below, tick (✓) the measures we can take to avoid infection from pets or insects.

1. Go near stray animals and play with them. ☐
2. Hold any unknown insect seen in or around our homes. ☐
3. Bring wild animals or insects home to be kept as pets. ☐
4. Play in clean areas such as playgrounds and parks. ☐
5. Wash your hands after playing with a pet or after petting it. ☐
6. Vaccinate your pets regularly and keep them away from stray animals. ☐

L Cross-curricular Learning Link

Use the pie chart given below and find out the food items eaten by your family members for that day and fill in the coloured parts of the chart with the names of the food items accordingly. Find out if it is a balanced diet and if not advise them on the benefits of taking a balanced diet.

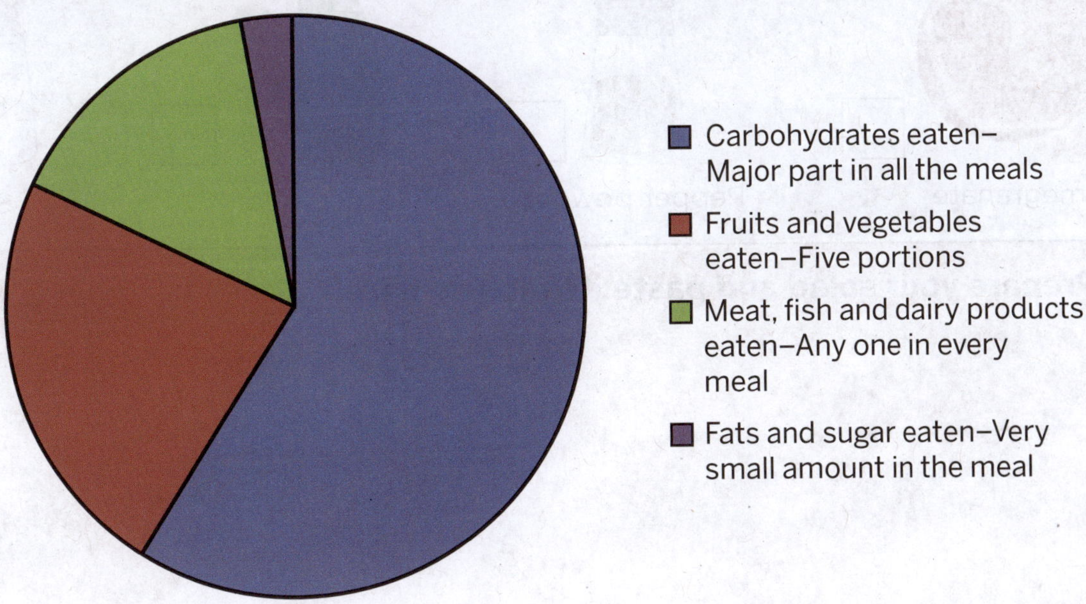

WORKSHEET

Salads are known to be rich sources of roughage, and they also contain vitamins and minerals. Tick the food items you would use to make your favourite salad.

Sprouted green gram ☐

Cauliflower ☐

Carrot ☐

Ginger ☐

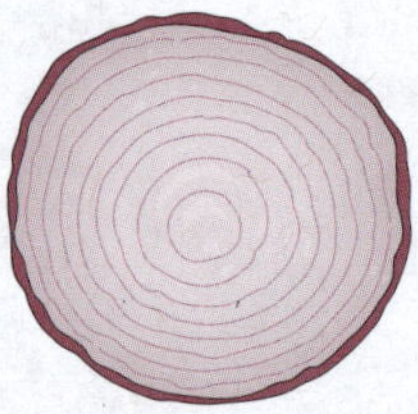
Onion ☐

Broccoli ☐

Cooked potatoes ☐

Coriander ☐

Mint ☐

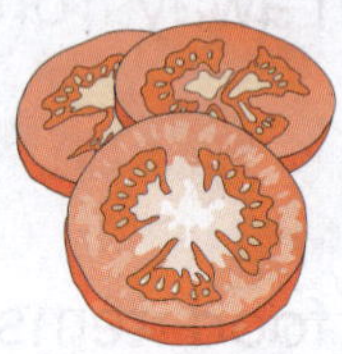
Tomato ☐

Cabbage ☐

Green chillies ☐

Pomegranate ☐

Pepper powder ☐

Cucumber ☐

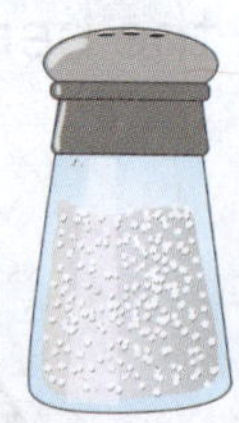
Salt ☐

Prepare your salad and paste its picture here!

5 Safety and First Aid

Learning Checklist

- Safety from fire
- First aid

Let's Get Started

Look at the pictures given alongside and tick (✔) the correct answer for the given questions.

1. What should you do while lighting firecrackers?
   - You should light the firecrackers on your own. ☐
   - You should wear only synthetic clothes while lighting crackers. ☐
   - You should light the firecrackers only in the presence of an adult. ☐

2. How do you cross the road?
   - Before crossing, first look to your right, then to your left and make sure that all the vehicles have stopped. Then cross the road. ☐
   - Always run while crossing the road so that you can reach the other end fast. ☐
   - As soon as you find a zebra crossing, run towards the other end without waiting for the signal. ☐

Accident

Accidents occur due to carelessness, when people do not follow safety rules. It is important to follow safety rules everywhere to avoid accidents.

You have learned in your previous classes about safety rules to be followed at home, in school, on the road and in playground.

Let us now learn how to be safe from fire and the first aid to be given for burns, animal bites, nosebleed, etc.

SAFETY FROM FIRE

A fire can be caused due to any of the following reasons:

- Carelessness when handling burning matchsticks and candles is one of the main causes for fire accidents.
- Carelessness while burning firecrackers is also a major cause of fire accidents during festivals.
- Electrical faults can cause fire outbursts.
- Carelessness during cooking and gas leaks in the kitchen can also cause fire.

Lighting firecrackers in the presence of an elder

To prevent fire accidents, we must follow some safety rules:

- We should be very careful while handling fire. Children must not handle fire on their own.
- Special care should be taken while lighting firecrackers. Make sure that you light firecrackers only in the presence of an elder.
- Never play with matchsticks or candle flame.
- Electrical appliances and wiring should be checked regularly for faults or wear and tear.
- Electrical faults should be fixed immediately.
- We should not wear nylon or synthetic clothes while lighting firecrackers or while cooking as they catch fire easily. Cotton or woollen clothes are safer.

Checking electrical appliances

- In case of a gas leak, open all doors and windows immediately. Do not strike a matchstick or handle any electric switch. Electric switches produce sparks when switched on or off. This may cause fire when it comes in contact with gas.
- Faulty gas pipes should be replaced.

Leakage of gas

Putting Out Fire

We can put out fire by the following ways:

- Fire may spread very quickly. You know that fire needs air to keep burning. One method of putting out fire is to cut off the air supply. This can be done by throwing mud or sand over the fire.
- In case of a major fire, inform the nearest fire station. Firefighters put out fire with the help of a fire engine. A small fire can be put out using a special instrument called the **fire extinguisher**.
- In case your clothes or someone else's clothes catch fire, do not run around. This will fan the flames and make them spread. Stop where you are; drop to the ground and roll on the ground with the hands covering your face. Rolling puts off the flames. You can also use water to put out fire or wrap yourself with a thick blanket or rug. But do not stay covered with a blanket for too long.

Fire engine

Firefighter putting out fire

Revise and recall

Circle the words that are related to fire.

Matches, campfire, pedestrian crossing, stove, traffic signal, bonfire, candles, smoke, crackers, knife, lighter, flame and sparks

FIRST AID

When a person gets injured, a doctor or a nurse may not be available at that particular place and time. In such cases, the injured person should be given first aid immediately. **First aid** is the first or immediate help given to an injured person before he or she is taken to a doctor.

First Aid for Cuts

Any cut or open wound on the skin should not be left unattended as germs can enter our body through them. Wash the cut with clean water. Wipe it and apply an antiseptic cream or lotion. If there is loss of blood, tie a clean cloth or use a band-aid over the cut. One must go to a doctor in case of deep cuts.

First Aid for Burns

In case of minor burns, keep the burnt area under cold running water for sometime or apply ice cubes immediately. This will cool the burnt area. Then apply an antiseptic cream on that area.

If the burn forms blisters[1], never try to prick them. Cover them with a soft cotton cloth. In case of severe burns, visit the nearby doctor immediately.

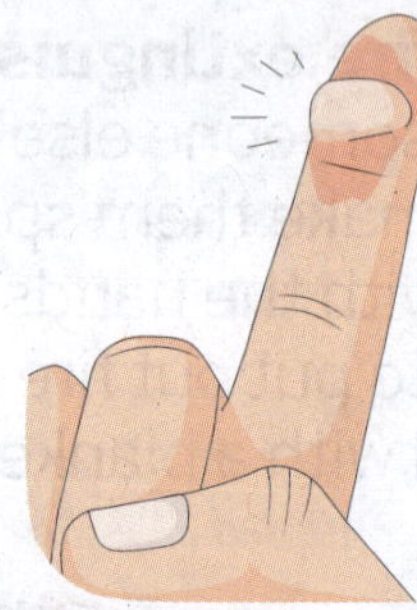

Blister

Cooling the burnt hand under cold running water

Science in Context

A solution of baking soda in water can be applied to the burnt area to get relief from the burning sensation.

Word Web

[1]**Blister**: a small bubble-like swelling on the skin filled with water
[2]**Sling**: a strap or cloth that is tied around the neck to support a broken arm
[3]**Splint**: a piece of hard material (wood or cardboard) used to hold a broken bone in place

First Aid for Fracture

A break or crack in the bone is called a **fracture**. When a bone gets fractured, there will be pain and swelling around the injured area. Ensure that the fractured part is not moved. For example, if there is a fracture in the arm, a cloth sling[2] made from a triangular piece of cloth can be used to support the arm. A splint[3] can also be used to give support to the broken bone. The injured person should be taken to the nearby hospital immediately.

Sometimes, the ankle or wrist can get twisted and cause pain. This is called a **sprain**. We should not move the sprained part. An elastic bandage can be wrapped around the sprained part.

A sling for a fractured arm

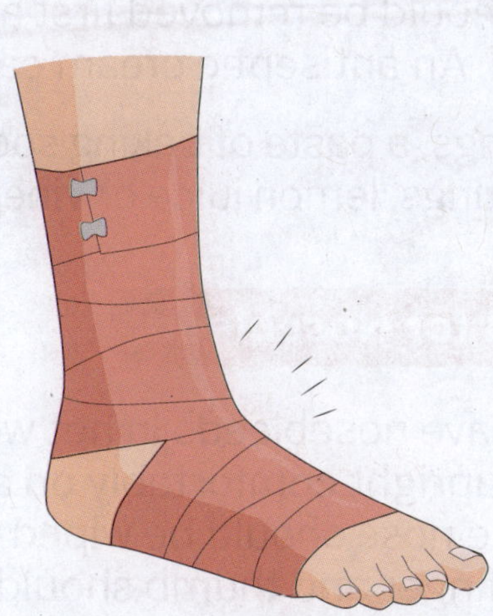

A bandage around a sprained ankle

First Aid for Animal Bite

The bites of stray animals like dogs, cats and monkeys can cause a disease called **rabies**. When these animals bite someone, they release saliva, containing germs, from their mouth. The saliva enters the victim's[1] body through the wound and causes rabies. So, in case of an animal bite, the wound should be washed with soap under running water. An antiseptic cream should be applied and the wound should be covered with a clean bandage. The victim should be taken to the nearby doctor immediately.

Science in Context

It is important to identify the animal that bit a person so that the nurse or doctor can treat the patient for rabies, if necessary.

Word Web

[1]**Victim**: a person who has been attacked or injured

First Aid for Snakebite

Snakes mostly hide in bushes. We need to be careful while playing in a garden near wild bushes. If a person is bitten by a poisonous snake, the poison enters in the victim's body. The poison mixes with blood and spreads quickly in the body. In case of a snakebite, the following first aid should be given:

- A tight bandage should be tied just above the wound so that the spread of poison in the body may slow down.
- The victim should be kept calm and taken to the nearby doctor immediately.

First Aid for Insect Sting

Sometimes, we may get stung by insects such as a bee or a wasp. In such cases, the sting should be removed first and then some ice cubes should be placed on the bitten area. An antiseptic cream can also be applied on the bitten area.

For bee stings, a paste of baking soda or toothpaste can be applied. For wasp stings, lemon juice or vinegar can be applied.

First Aid for Nosebleed

First aid for nosebleed

One may have nosebleeds in hot weather. In such cases, one should sit upright comfortably on a chair and lean forward slightly. The nose should be wiped with a clean handkerchief. The index finger and thumb should be used to pinch the nose. The person should breathe through the mouth and not through the nose.

If the nosebleed does not stop, the person should be taken to a doctor.

Let's Say Together!

Firecrackers Blister Sling Sprain Fracture Rabies

Keywords

First aid: The first or immediate help given to an injured person before he or she is taken to a doctor

Fracture: A break or crack in a bone

Sprain: An injury by a sudden twisting of joints that causes swelling and pain

Rabies: A disease caused by animal bites

Summary

1. Accidents occur due to carelessness and when people do not follow safety rules.
2. In case of a gas leak, open all doors and windows immediately. Do not strike a matchstick or handle any electric switch.
3. In case of a major fire, inform the nearest fire station.
4. First aid is the first or immediate help given to an injured person before he or she is taken to a doctor.
5. The bites of dogs, cats and monkeys can cause a disease called rabies.
6. In case of a snakebite, tie a tight bandage just above the wound so that the spread of poison in the body may slow down.

EXERCISE

A Choose the correct option for each of the following.

1. Fire needs ______________ to keep burning.
   (a) air (b) water (c) smoke (d) light
2. Nylon or synthetic clothes should not be worn while bursting firecrackers or while
   (a) going out (b) sitting (c) cooking (d) playing
3. In case of minor burns, keep the burnt area under ______________.
   (a) hot water (b) cold running water (c) boiling water (d) salt solution
4. Rabies is caused by
   (a) snakebites (b) animal bites (c) gas leak (d) fracture
5. Vinegar or lemon juice can be applied for
   (a) bee stings (b) burns (c) wasp stings (d) cuts

B Fill in the blanks using the correct words.

1. ______________ can cause fire outbursts. (Fire engines/Electrical faults)
2. A small fire can be put out using a fire ______________. (engine/extinguisher)
3. ______________ clothes do not catch fire easily. (Cotton/Nylon)
4. ______________ down on the ground puts off the flames on your clothes. (Bending/Rolling)
5. An elastic bandage can be wrapped around a ______________ part. (fractured/sprained)

C Match the following.

Column A	Column B
(a) Band-aid	(1) Burns
(b) Blisters	(2) For fracture
(c) Splint	(3) Hot weather
(d) Ice cubes	(4) For cuts
(e) Nosebleed	(5) Cool the burnt area

D **Name the following.**

1. Mud or sand is thrown over it to put it out ______________________
2. Firefighters put out fire with the help of this ______________________
3. A cloth that is tied around the neck to support a broken arm ______________________
4. A disease that can be caused by an animal bite ______________________
5. A person who has been attacked or injured ______________________

E **Answer the following questions briefly.**

1. How are fires caused?
2. What should be done in case of a gas leak?
3. What is a fire extinguisher?
4. What first aid should be given for a fracture?
5. What is a sprain? What first aid should be given for a sprain?

F **Answer the following questions.**

1. Write any three safety rules to be followed to prevent fire accidents.
2. Write the various methods of putting out fire.
3. Define first aid. What first aid can be given in case of burns?
4. What first aid can be given to a victim in case of an animal bite?

G **HOTS (Higher Order Thinking Skills)**

1. You are playing with your friend in a garden. Suddenly your friend gets stung by a bee. What will you do? (Hint: First aid for insect sting)
2. A person whose clothes catch fire should not run around. Why? (Hint: Air helps fire to burn)

H **Observe the pictures given below and write the type of accidents/injury shown in each.**

____

____

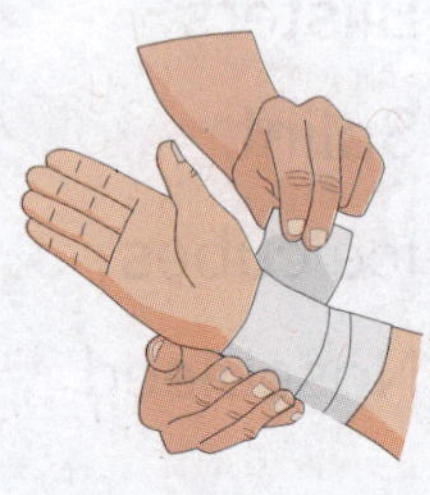
____

John has been bitten by a dog. What first aid should he take? Number the steps given below in the correct order so that John can follow them easily.

Sentences	Number
If there is bleeding, he should press the wound and try to control it.	
He should go to a doctor immediately.	
John should wash his wound thoroughly in running water with soap.	
He should cover the wound with a clean bandage.	
He should inform an adult about what animal bit him.	

Project

Fumy, the fire smarter, has been selected for campaigning on fire safety. Join him for the campaign by making a poster on a few tips on safety from fire.

Life Skills

Whom do you call if there is an emergency? Write down the phone numbers of the following. Always have these telephone numbers with you.

1. Your parents ____________________
2. Police ____________________
3. Ambulance ____________________
4. Fire station ____________________
5. Your uncle or any other person whom you trust (this is required in case your parents' numbers are not reachable) ____________________
6. Family doctor ____________________

6 Air and Water

Learning Checklist

- Layers of the atmosphere
- Composition of air
- Properties of air
- Water
- Purification of drinking water

Let's Get Started

Rekha and Smita are sisters. They made a plan to celebrate their grandfather's birthday and bought balloons to decorate the house.

Look at the pictures and answer the question given below each.

What is Rekha blowing into the balloon? ______________________

Why did Smita's balloon burst? ______________________

Air and water are two of our basic necessities. In this chapter, we will learn about the composition and properties of air, and the ways to purify water.

LAYERS OF THE ATMOSPHERE

The Earth is surrounded by a thick envelope of air called the **atmosphere**. In order to study the atmosphere, scientists have divided it into five layers.

Troposphere: This is the first layer of the atmosphere, just above to the ground. Troposphere is where we live in and where most aeroplanes fly. The Weather changes take place in this layer.

Stratosphere: This is the second layer of the atmosphere. The ozone layer is a part of stratosphere. The ozone layer absorbs harmful radiations from the sun and stops them from reaching Earth's surface.

Jet planes fly in the stratosphere.

Mesosphere: This is the third layer of the atmosphere from the Earth's surface. This is the coldest among all the atmospheric layers.

Meteors[1] burn up in this layer while entering the Earth's surface. From the Earth, we see them as shooting stars.

Thermosphere: This is the fourth layer of the atmosphere from the Earth's surface. Space shuttles fly in this layer.

Exosphere: This is the uppermost layer of the atmosphere. There is no air beyond this.

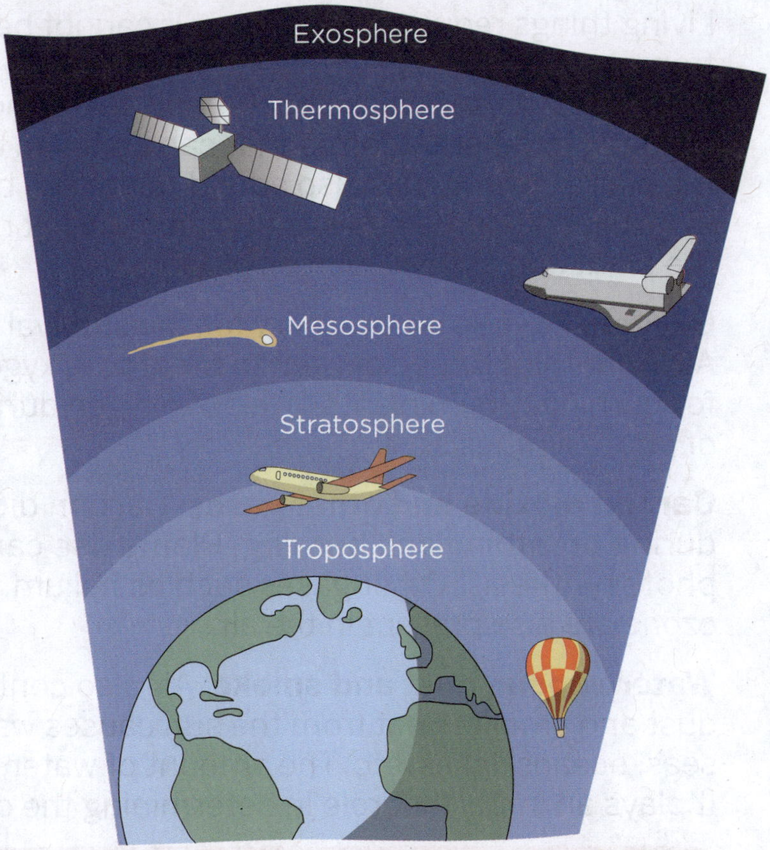

Layers of the atmosphere (not to scale)

Word Web

[1]**Meteors**: rocks from outer space that enter the Earth's atmosphere

The atmosphere is useful to us in many ways:

- It contains oxygen, which is essential for the survival of living things. It also contains carbon dioxide, which is required for photosynthesis.
- It helps to maintain the right temperature for the survival of living things.
- It has the ozone layer, which protects us from the harmful rays of the sun.

COMPOSITION OF AIR

Air is a mixture of gases. It mainly consists of nitrogen (78%), oxygen (21%) and other components such as carbon dioxide and other gases (1%). Air also contains small quantities of water vapour, dust and smoke.

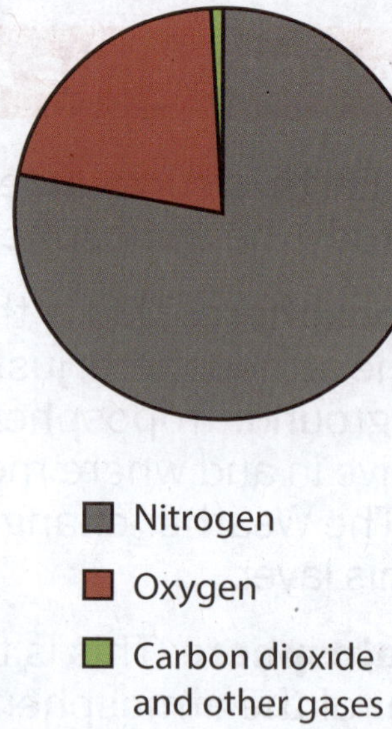

Nitrogen: It is the most abundant gas in the atmosphere. Living things require nitrogen but it cannot be taken directly from the air. Plants get nitrogen with the help of bacteria present in the soil. These bacteria fix the atmospheric nitrogen to different forms that can be taken up by the plants. Nitrogen is also added to the soil by means of fertilizers. Human beings and animals get nitrogen from the plant and animal sources of food.

Oxygen: Oxygen is necessary for the survival of living things. All living things need oxygen to breathe. Oxygen is also needed for burning. Green plants give out oxygen during the process of photosynthesis.

Carbon dioxide and other gases: Carbon dioxide is released into the atmosphere during breathing and burning. Plants use carbon dioxide for the process of photosynthesis. Other gases such as helium, neon, hydrogen, argon, krypton and ozone are also present in the air.

Water vapour, dust and smoke: Air also contains a small amount of water vapour, dust and smoke. Heat from the sun causes water to evaporate from the surface of seas, oceans, lakes, etc. The amount of water present in the air is called **humidity**. It plays an important role in determining the daily weather.

Activity 1

Aim: To show that oxygen is necessary for burning

Materials required: Two candles and a glass jar

Procedure:

1. Place the candles on a table and light them.
2. After a while invert the glass jar over one of the candles.

Observation: The candle that is exposed to air continues to burn, whereas the candle covered by the jar burns for a short while and then stops burning.

Reason: The flame of the candle in the jar goes off because covering the candle cuts off the oxygen supply.

Conclusion: This shows that oxygen is essential for burning.

Air is used in burning, breathing and also in a process called **rusting**. Objects made of iron get covered with a reddish brown substance called **rust**. This happens when iron objects come in contact with air and the water vapour present in the air.

PROPERTIES OF AIR

We know that air has weight and it occupies space. Let us do these simple activities to understand these properties of air.

Activity 2

Aim: To show that air has weight

Materials required: Two balloons, a needle, two pieces of thread and a scale

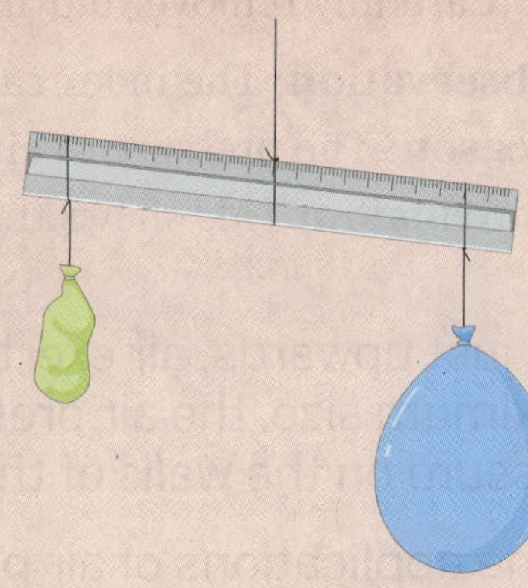

Procedure:

1. Blow air into both the balloons.
2. Tie the balloons with the help of threads to either side of the scale.
3. Lift the scale. Observe how the scale is positioned.
4. Now, burst one of the balloons with a needle.
5. Observe what happens.

Observation: On lifting the scale, it stays horizontal. But when one of the balloons is burst, the scale tilts towards the balloon that is still intact.

Reason: This happens because the intact balloon has air inside it, which makes it heavier. Hence, the scale tilts.

Conclusion: This shows that air has weight.

Activity 3

Aim: To show that air occupies space

Material required: A balloon

Procedure:

1. Hold the balloon in your mouth and blow into it.
2. Observe what happens.

Observation: The balloon increases in size when you blow into it.

Reason: The air we blow into the balloon occupies space. Hence, the balloon increases in size.

Conclusion: This shows that air occupies space.

Because of its weight, air pushes everything around it in all the directions. In other words, we say that air exerts pressure. This pressure is called the **air pressure**. Let us do a simple activity to show that air exerts pressure.

Activity 4

Aim: To show that air exerts pressure

Materials required: Plastic cup and index card (cardboard)

Note: It might be a good idea to try this over a sink.

Procedure:

1. Fill the plastic cup with water. Place the index card on top of the cup.
2. Hold the index card tightly against the mouth of the plastic cup and turn the cup upside down.
3. Carefully remove your hand holding the index card in place.

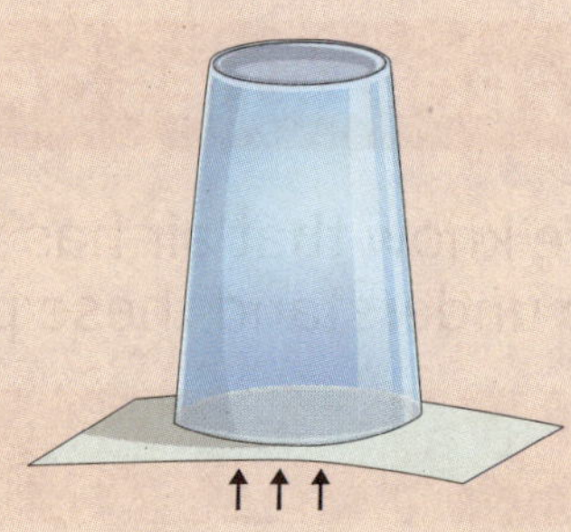

Observation: The index card does not fall.

Reason: The pressure of air outside the plastic cup is pushing the card up.

Conclusion: This shows that air exerts pressure on objects.

Not just upwards, air exerts pressure on all sides. When a balloon is blown over its maximum size, the air pressure increases. At some point, the air exerts too much pressure on the walls of the balloon and it bursts.

Some applications of air pressure in everyday life are given below:

Drinking through a straw

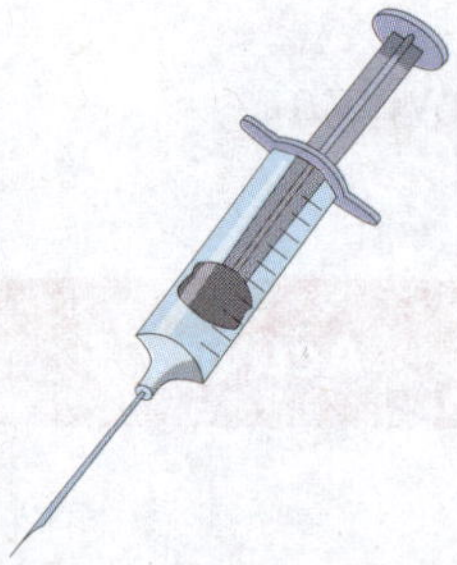

Filling up injection syringes

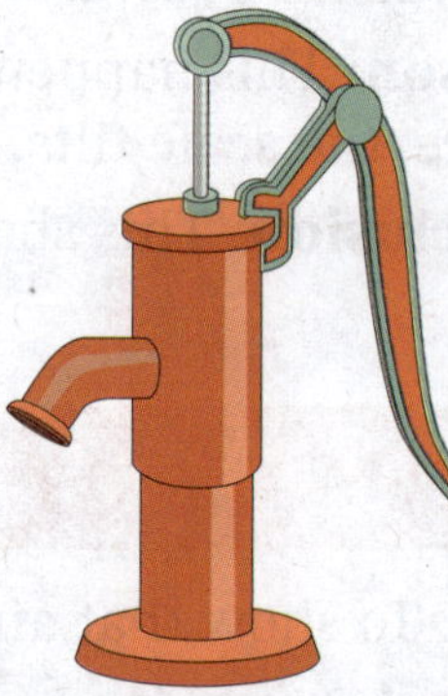

Getting water from a hand pump

Using a vacuum cleaner

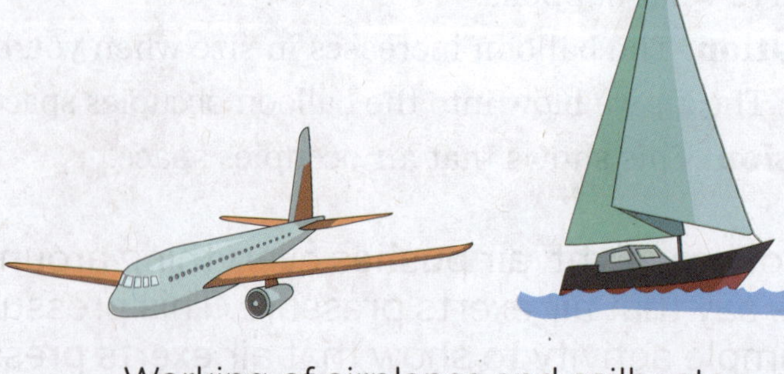

Working of airplanes and sailboats

Revise and recall

Fill in the blanks using the correct words.

1. ______________ is added to the soil by means of fertilizers. (Nitrogen/Argon)
2. All living things need ______________ to breathe. (oxygen/nitrogen)
3. The layer of atmosphere where weather changes take place is ______________. (troposphere/stratosphere)
4. Space shuttles fly in ______________. (exosphere/thermosphere)
5. Air exerts ______________ on objects. (weight/pressure)

WATER

Like air, water is also essential for the survival of living things. We need water for drinking, cooking, bathing, etc. It is very important that we drink clean and pure water. Rainwater, rivers, ponds, lakes and wells are some main sources of water. Water from these sources contains substances such as sand, dirt and germs, which make the water impure and thus unfit for use.

Water needs to be purified to make it fit for human consumption[1]. Water that is fit for human consumption is called **potable water**.

Things that make the water impure are called **impurities**. Impurities can be soluble or insoluble. **Insoluble impurities** are the ones that do not dissolve in water. **Soluble impurities** are the ones that dissolve in water.

Word Web

[1]**Consumption**: eating or drinking

Removing Insoluble Impurities from Water

Insoluble impurities such as sand and mud can be removed by sedimentation and decantation, and filtration methods.

Activity 5

Aim: To understand the processes of sedimentation and decantation

Materials required: Two beakers, glass rod, sand and water

Procedure:

1. Take the beaker containing the mixture of sand and water.
2. Use the glass rod to mix the contents. Leave the beaker undisturbed for some time.
3. The heavier sand particles will settle at the bottom of the beaker as shown in the figure given alongside. This is called **sedimentation**.
4. Without disturbing the sediment of sand, carefully drain out only the clear water into another beaker as shown in the figure. This is called **decantation**.

Sedimentation

Decantation

The process of separating solid particles from a liquid medium using a filter paper is known as **filtration**.

Activity 6

Aim: To understand the process of filtration

Materials required: Two beakers, glass rod, filter paper, funnel, soil and water

Procedure:

1. Take some soil in a beaker and add some water to it. Use a glass rod to mix the contents.
2. Take a filter paper and make a cone as shown in the adjacent pictures.
3. Place the paper cone in the funnel. Now, place the funnel on the mouth of another beaker.
4. Gently pour the mixture of soil and water over the filter paper.
5. You will notice clear water dripping through the funnel into the beaker, while soil particles get retained in the filter paper.
6. The clear water that gets collected in the beaker after passing through the filter paper is called the filtrate.

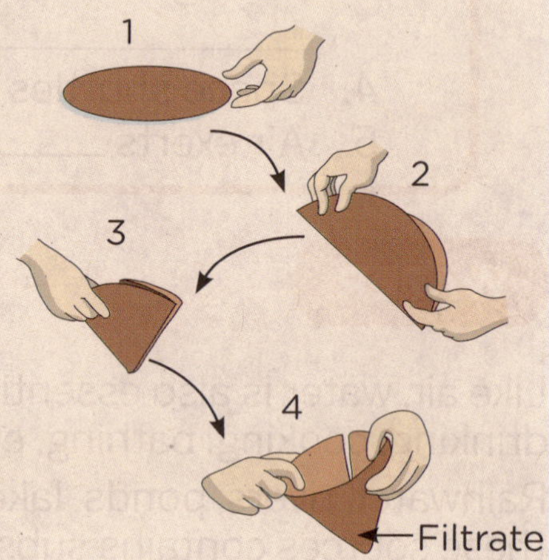

Applications of filtration in daily life

- Vacuum cleaners have filters to trap dust particles.
- Water taps usually have filters to purify water.
- Coffee makers have filters to keep coffee grounds out of coffee.
- Air conditioners have filter to trap the dust particles in the air.

Vacuum cleaner

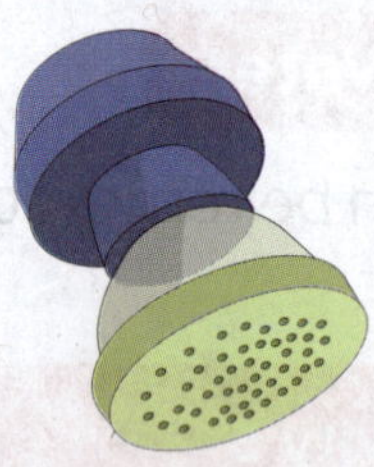

Water tap

Coffee filter

Air conditioner

Removing Soluble Impurities from Water

The soluble impurities in water can be removed by evaporation and distillation methods.

Common salt (otherwise called sodium chloride) is extracted from seawater by the process of evaporation under the sun. Let us perform an activity to understand this process.

Activity 7

Aim: To understand the process of obtaining salt from seawater by evaporation

Materials required: A shallow dish, saltwater and Bunsen burner

Procedure:

1. Set up the apparatus as shown in the picture.
2. Add saltwater to the shallow dish and heat it till all the water evaporates.

Observation: As the water evaporates, we can see salt getting deposited along the inner walls of the dish.

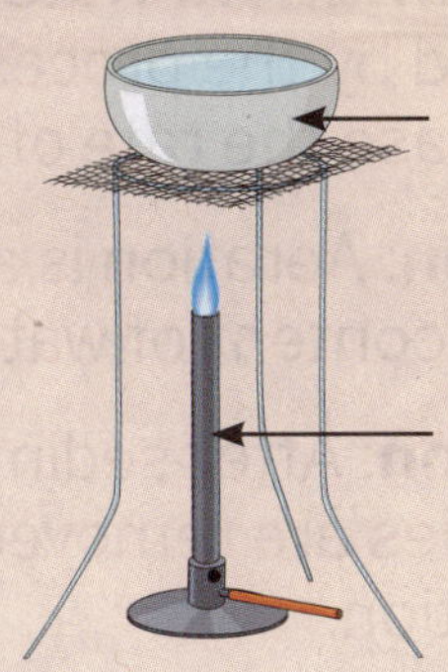

Distillation is a process that can be used to separate a pure liquid from a mixture of liquids due to the difference in their boiling points. The laboratory apparatus for distillation is shown below:

The process involves the following steps:

1. Impure water is heated in a round-bottom flask.
2. Water evaporates to form water vapour and the impurities are left behind.
3. Water vapour passes through the condenser and cools down to form water.
4. Pure water is collected in a separate container.

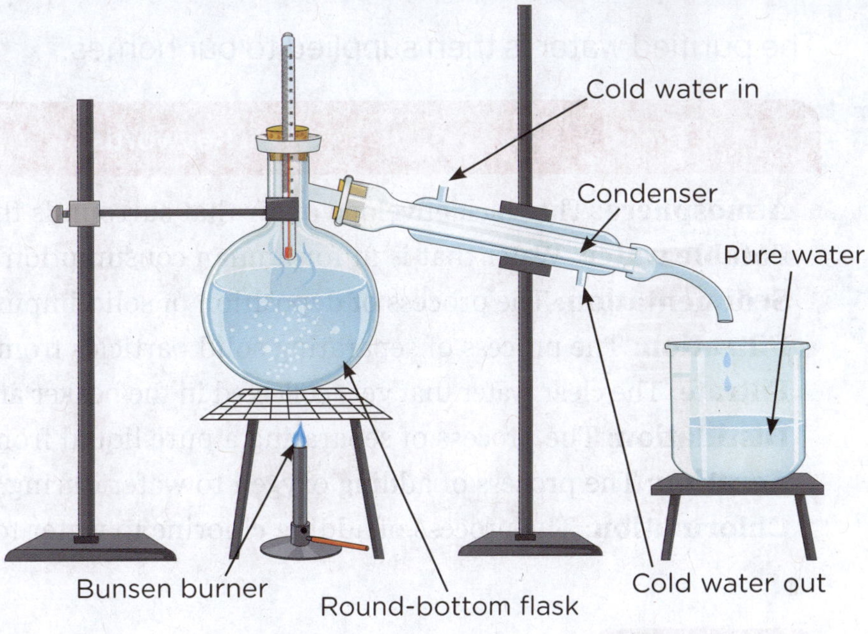

Apparatus for distillation of water

Distilled water is considered to be the purest form of water as it does not contain any impurities. It is used in science laboratories, car batteries and medicines.

PURIFICATION OF DRINKING WATER

Water purification is the process of removing germs and other harmful substances from water to make it fit for consumption.

At home, we can purify water by boiling it for about 10 minutes. It is the simplest way of purifying water.

The drinking water that is supplied to our homes is first treated to make it fit for drinking. The following steps are used in the process:

Sedimentation: Water is collected in large tanks where a lot of impurities are removed by the process of sedimentation. Chemicals such as alum are added to increase the rate of sedimentation.

Aeration: Aeration is a process in which the oxygen content of water is increased.

Filtration: After sedimentation, finer impurities are removed by the process of filtration.

Chlorination: It is the process of adding chlorine to water to kill harmful germs found in water. Chlorination can be done at any point during the water purification process.

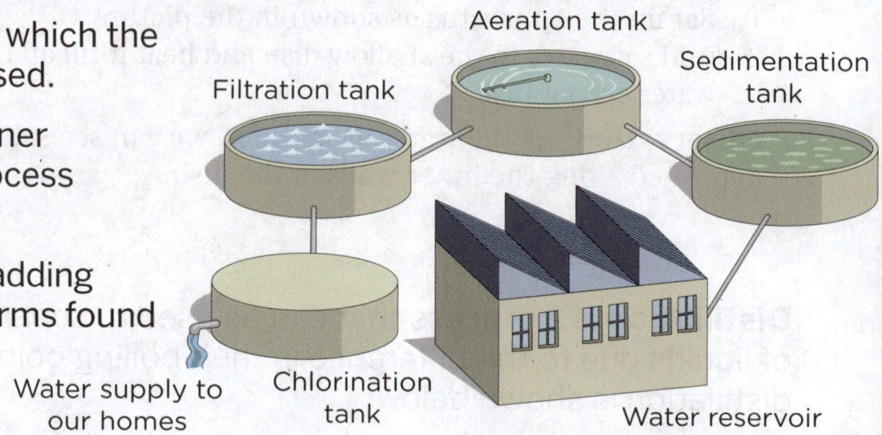

Purification of water

The purified water is then supplied to our homes.

Keywords

Atmosphere: The thick envelope of air that surrounds the Earth

Potable water: Water that is fit for human consumption

Sedimentation: The process of deposition of solid impurities present in a liquid

Filtration: The process of separating solid particles from a liquid using a filter paper

Filtrate: The clear water that gets collected in the beaker after passing through the filter paper

Distillation: The process of separating a pure liquid from a mixture of liquids

Aeration: The process of adding oxygen to water during the water purification process

Chlorination: The process of adding chlorine to water to kill harmful germs in water

Summary

1. Air is a mixture of gases.
2. Air has weight, occupies space and exerts pressure on all objects.
3. The thick envelope of air that surrounds the Earth is called the atmosphere.
4. The five layers of atmosphere are troposphere, stratosphere, mesosphere, thermosphere and exosphere.
5. Water can be purified by sedimentation, decantation, filtration and distillation methods.

EXERCISE

A Choose the correct option for each of the following.

1. The most abundant gas in the atmosphere is
   (a) Nitrogen (b) Oxygen (c) Carbon dioxide (d) Water vapour
2. __________ is the gas that is used in photosynthesis.
   (a) Oxygen (b) Nitrogen (c) Carbon dioxide (d) Argon
3. Where is the ozone layer found?
   (a) Troposphere (b) Stratosphere (c) Mesosphere (d) Exosphere
4. In water purification, alum is added during the process of
   (a) sedimentation (b) filtration (c) aeration (d) chlorination
5. Which of the following chemicals is used to kill germs during water purification?
   (a) Fertilizer (b) Alum (c) Chlorine (d) Argon

B State whether the following statements are true (T) or false (F).

1. Air exerts pressure always in the upward direction. ☐
2. Plants cannot use nitrogen directly from the atmosphere. ☐
3. Stratosphere is the coldest among all the atmospheric layers. ☐
4. Soluble impurities do not dissolve in water. ☐
5. Decanted water is the purest form of water. ☐

C Fill in the blanks with the correct words.

1. __________ is responsible for rusting. (Water vapour/Dust)
2. The Earth is surrounded by a thick envelope of air called the __________. (atmosphere/ionosphere)
3. The uppermost layer of the atmosphere is __________. (troposphere/exosphere)
4. __________ removes soluble impurities from water. (Filtration/Distillation)
5. __________ is used to increase the rate of sedimentation. (Chlorine/Alum)

D Give two examples for each of the following.

1. Properties of air __________ __________
2. Gases present in the air __________ __________
3. Applications of air pressure __________ __________

4. Processes for removing insoluble impurities from water ________ ________

5. Uses of distilled water ________ ________

E Answer the following questions briefly.

1. What is air pressure?
2. Name the different layers of the atmosphere.
3. How does the ozone layer protect the Earth?
4. Why should we drink purified water?
5. Write any two applications of filtration method in our daily life.

F Answer the following questions.

1. What is the difference between troposphere and stratosphere? Explain.
2. How will you prove that air exerts pressure? Explain with the help of a diagram.
3. Define distillation and list the steps involved in this process.
4. Explain the steps involved in water purification before its supply to our homes.

G HOTS (Higher Order Thinking Skills)

1. You have to separate orange juice from its pulp. Which process out of these would you use: sedimentation and decantation or filtration? Why?
2. Iron grills in our windows and gates are painted.
(Hint: Painting does not expose them to air)

H Label the parts of the distillation apparatus given below.

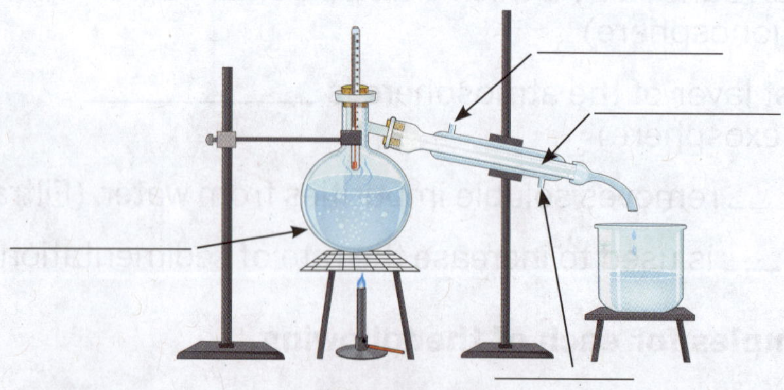

I Project

1. On a chart paper, draw the process of water purification. Explain each step.
2. Nowadays, a lot of water purifiers are available in the market. With the help of an elder, find out how they work, collect pictures and prepare a report.

Label the different layers of the atmosphere.

Life Skills

Water is a precious resource. Let us save every drop we can.

Turn off taps while brushing your teeth, soaping clothes and scrubbing vessels.

Fixing leaking taps does not take much, but it can save a significant amount of water.

Using a bucket instead of a shower to bathe can reduce your water consumption by 80 per cent!

Energy-star-rated dishwashers and washing machines use less water.

Low-flow showerheads and toilets with smaller flush tanks use less water.

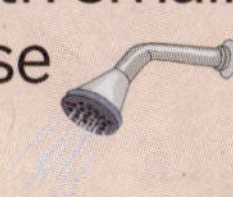

To recycle water on your own, do simple things like using the water in which vegetables have been washed to water plants.

Write three other ways to save water.

1. ______________________________

2. ______________________________

3. ______________________________

7 Earth, Sun and Moon

Learning Checklist

- The solar system
- Sun
- Earth
- Moon
- Eclipses

Let's Get Started

Identify the objects shown in the pictures below that we normally see in the sky.

Objects seen in the sky are called **celestial objects**. The sun and the other celestial objects that revolve around it form the solar system.

Let us learn more about the solar system, the sun, the Earth and the moon.

THE SOLAR SYSTEM

The sun and the eight planets that revolve around the sun together form the **solar system**. Other than the planets, satellites[1] of various planets, comets and asteroids are also part of the solar system.

> **Word Web**
>
> [1]**Satellite**: a celestial object that moves around a planet

The sun is at the centre of the solar system. There are eight planets revolving around the sun. These planets are Mercury, Venus, Earth, Mars, Jupiter, Saturn, Uranus and Neptune. They revolve around the sun in fixed paths called **orbits**.

Mercury is the smallest planet and is also the closest planet to the sun. Venus is the hottest and brightest planet. Earth is also called the blue planet. It is the only planet where life is known to exist. Mars is also called the red planet. Jupiter is the largest planet of the solar system. Saturn is the second largest planet after Jupiter. It has rings around it. Saturn is so light that if there was a tub of water big enough to hold it, it would float in it. Uranus is the coldest planet of the solar system. Neptune is the planet farthest from the sun.

Science in Context
Pluto, which was earlier considered as the ninth planet, is now known as a 'dwarf' planet.

The solar system (not to scale)

SUN

The sun is a star that is situated at the centre of our solar system. It is a huge ball of burning gases. Its size is about 109 times that of the Earth. Its mass is about 330,000 times that of the Earth. It is almost 99.86% of the total mass of the solar system.

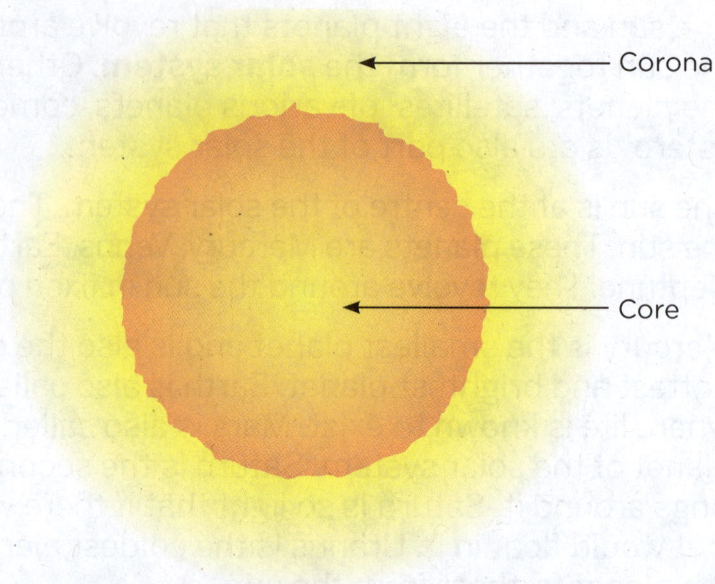

The Sun

Hydrogen and helium gases are the major constituents of the sun. The sun has a thin outer layer called the **corona** and a central part called the **core**.

In the core of the sun, hydrogen gas continuously gets converted into helium. A very large amount of heat and light energy is released during this process. Energy released by the sun is called the **solar energy**. The temperature on the surface of the sun is about 6000°C. All living things on the Earth ultimately depend on the solar energy for their existence.

EARTH

The Earth is the third planet from the sun. It is surrounded by an envelope of air known as the **atmosphere**. The atmosphere of the Earth protects us from the harmful radiations of the sun. The interior part of the Earth is divided into three main layers: crust, mantle and core.

Crust is the outermost layer of the Earth. Its thickness ranges from about 5 km below the ocean floor to about 70 km below the land. The continents and the oceans are on this layer. It is made up of rocks and contains all the things essential for life.

Mantle is the middle layer of the Earth lying between the crust and the core. The thickness of the mantle extends up to 2,890 km. The outer region of the mantle consists of solid rocks. The inner regions of the mantle consist of molten rocks. During volcanic eruptions, molten magma comes out from this layer to the surface of the Earth.

Core is the innermost layer of the Earth. It is mainly composed of iron and nickel. The outer core is in a molten liquid state. It extends to about 3,400 km. The inner core is a solid sphere of mostly iron situated about 1,220 km from the outer core.

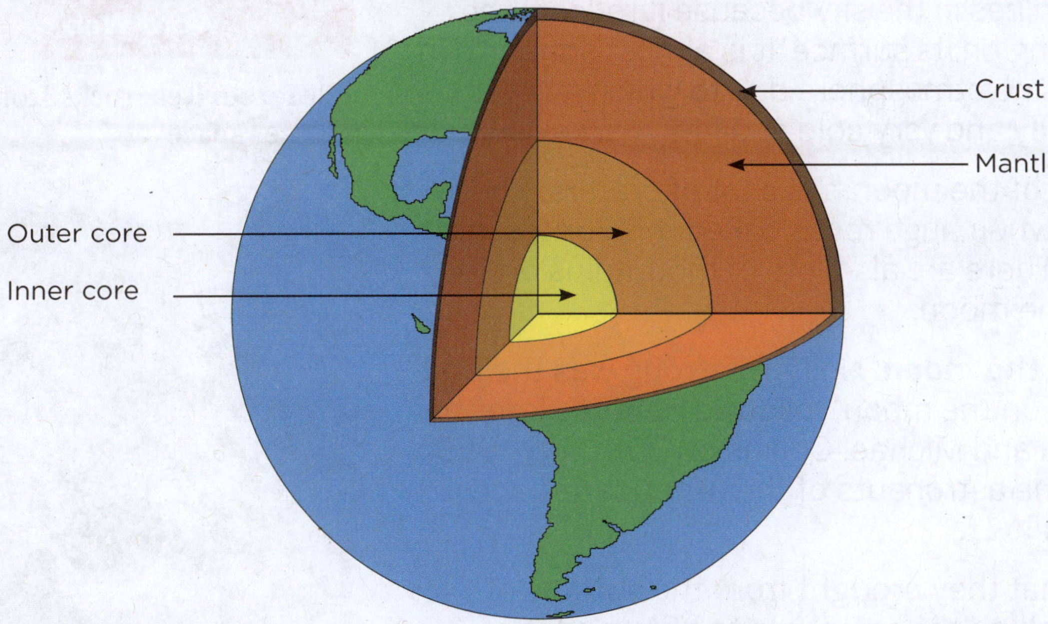

Internal structure of the Earth

Revise and recall

Fill in the blanks with the correct words.

1. The outer thin layer of the sun is called the _______________ . (core/corona)
2. In the core of the sun, _________ gets converted into helium. (hydrogen/nitrogen)
3. The interior part of the Earth is divided into _______________ main layers. (three/five)
4. The middle layer of the Earth is called the _______________ . (mantle/crust)
5. The major element in the inner core is _______________ . (iron/gold)

MOON

An object that revolves around a planet is called a **satellite**. Satellites are of two types: natural satellites and artificial satellites. Naturally existing objects revolving around a planet are called **natural satellites**. Some satellites of the Earth are man-made satellites. They are called **artificial satellites**.

The moon is the natural satellite of the Earth. It revolves around the Earth in an orbit at a distance of about 384,400 km from the Earth. The moon has no light of its own.

The moon shines in the sky because it reflects the sunlight falling on its surface. It is almost spherical in shape. It has no atmosphere due to which it is very hot during the day and very cold at night.

The surface of the moon has a lot of craters. These are formed when huge rocks called meteorites[1] hit its surface. There are also lots of mountains on the surface of the moon.

Landing on the moon: Neil Armstrong was the first man to land on the moon, followed minutes later by Edwin Aldrin and Michael Collins on 20th July 1969. They were the astronauts of the American space mission Apollo 11.

The rocks that they brought from the surface of the moon helped the scientists to obtain a lot of information about the moon.

There were some unmanned missions to the moon which obtained a lot of information about the surface of the moon through photographs.

Word Web

[1]**Meteorites**: rocks from outer space

Surface of the moon

Astronaut on the moon

Footprint on the moon

Chandrayaan Mission

Chandrayaan-1 was India's first unmanned mission to the moon. It was launched on 22 October 2008 from Satish Dhawan Space Centre, Sriharikota, Andhra Pradesh. The mission helped to obtain a lot of information about the moon.

Phases of the Moon

When we look at the moon, it appears to change its shape every night. Sometimes it is a round ball, sometimes C-shaped and sometimes it disappears completely. This happens because we see only that part of the moon that reflects the sun's light as it moves around the Earth in its orbit. These varied shapes of the moon are called the **phases of the moon**.

ECLIPSES

An **eclipse** is a natural phenomenon in which one celestial object (or a portion of it) becomes temporarily invisible as it is hidden by the shadow of another celestial object. It happens when three celestial objects are in a straight line.

The revolution of the Earth around the sun and the revolution of the moon around the Earth cause lunar and solar eclipses.

Lunar Eclipse

When the Earth comes between the sun and the moon, the shadow of the Earth falls on the moon. This causes the lunar eclipse.

Lunar eclipse can be **partial** in which a part of the moon may be visible or **total** in which the moon is not visible at all. The type and duration of a lunar eclipse depends on the position of the moon in the orbit.

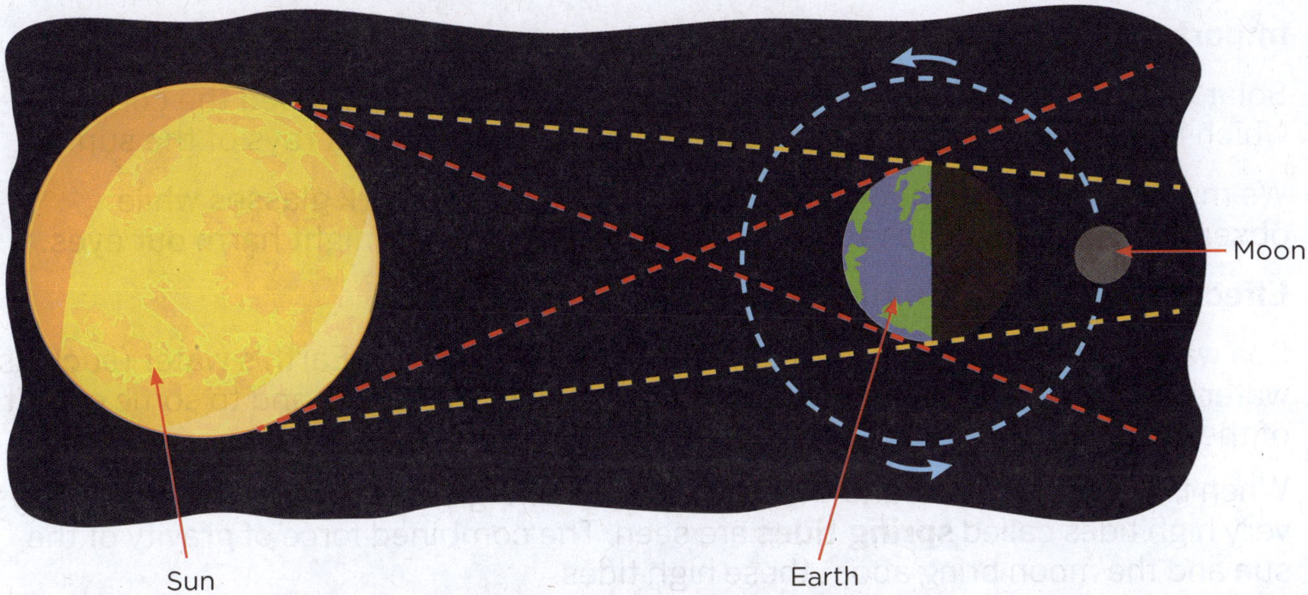

Lunar eclipse

Solar Eclipse

Solar eclipse occurs when the moon comes between the sun and the Earth, and the shadow of the moon falls on the Earth.

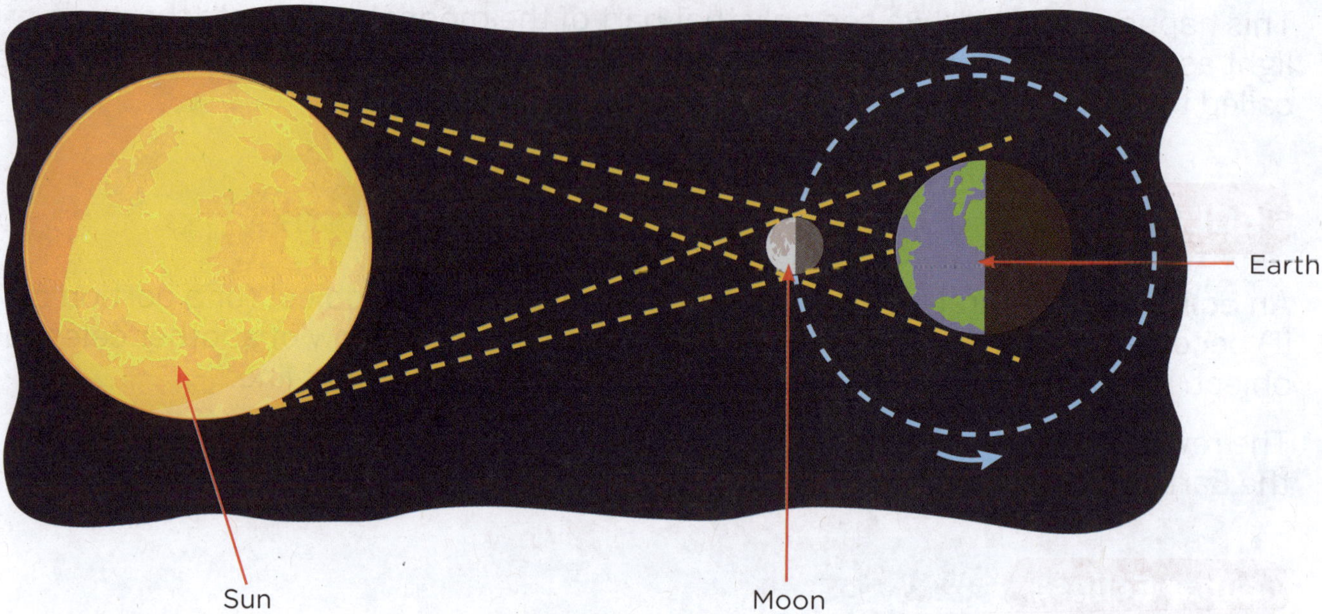

Solar eclipse

If we observe from an area on the Earth where the dark inner shadow of the moon falls, we can see a **total solar eclipse**. If we observe from an area on the Earth where the faint outer shadow of the moon falls, we can see a **partial solar eclipse**. If only a bright ring is seen around a central eclipsed area, then we see a ring-shaped eclipse or **annular eclipse**.

Importance of the solar eclipse

Solar eclipse enables us to study the outer layer of the sun called the corona, which is otherwise very difficult to observe due to the bright rays of the sun.

We must always remember to wear UV rays-protected dark glasses while observing the solar eclipse as the bright light of the sun might harm our eyes.

Effect of the eclipse on the Earth

Sea water rises and falls twice a day. The rise and fall of the Earth's water (such as water in the oceans) due to the gravitational pull of the moon, and to some extent of the sun, is called a **tide**.

When the sun, the moon and the Earth are in a straight line (during new moon days), very high tides called **spring tides** are seen. The combined force of gravity of the sun and the moon bring about these high tides.

When the sun and the moon are at a right angle to each other, we get very low tides called **neap tides**. They usually occur twice a month during the first and the third quarter of the moon.

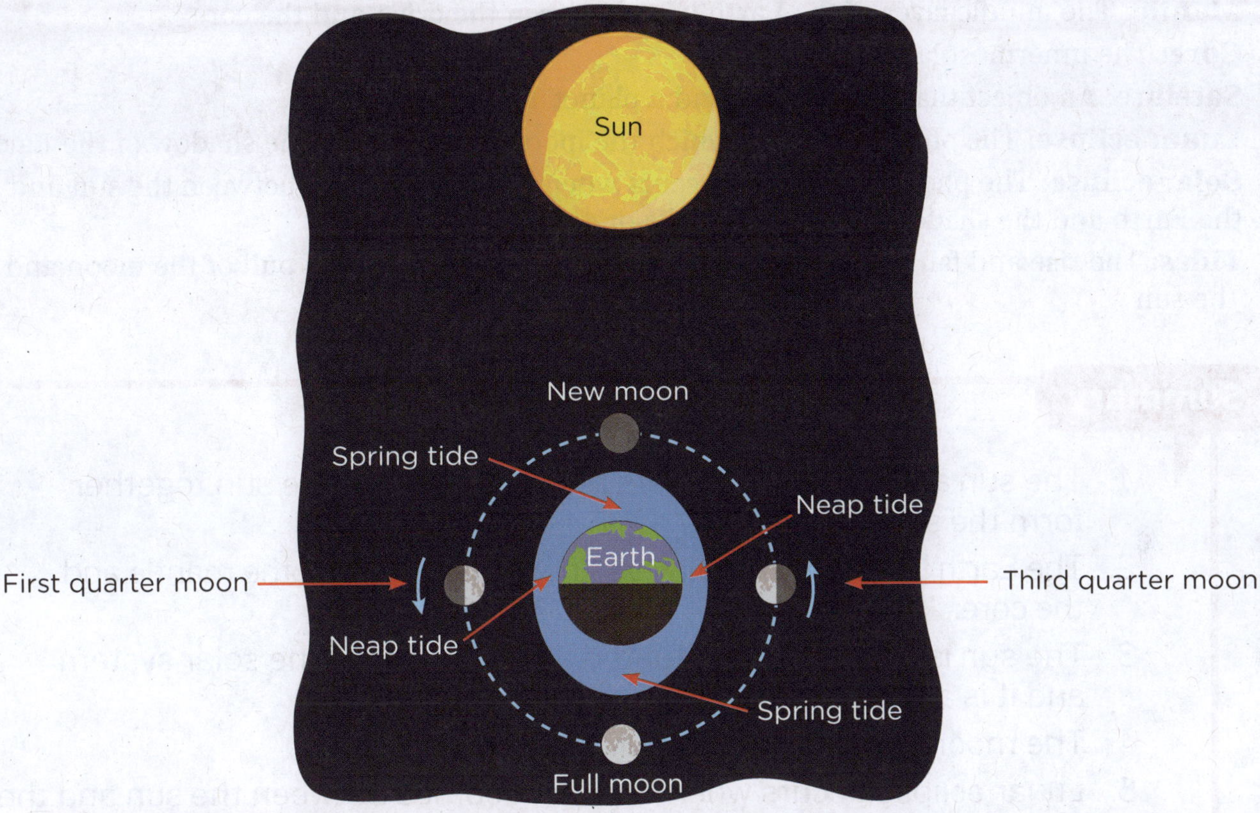

Spring and neap tides

Revise and recall

Match the following.

(a) Moon	(1) First man to land on the moon
(b) Neil Armstrong	(2) Shadow of the moon falls on the Earth
(c) Chandrayaan -1	(3) The sun and the moon at right angle
(d) Solar eclipse	(4) India's unmanned mission to the moon
(e) Neap tide	(5) The natural satellite of the Earth

Let's Say Together

Corona	Crust	Mantle	Satellite
Solar eclipse	Lunar eclipse	Spring tide	Neap tide

Keywords

Crust: The outermost layer of the Earth

Mantle: The middle layer of the Earth lying between the crust and the core

Core: The innermost layer of the Earth

Satellite: An object that revolves around a planet

Lunar eclipse: The phenomenon in which the moon is covered by the shadow of the Earth

Solar eclipse: The phenomenon that occurs when the moon comes between the sun and the Earth and the shadow of the moon falls on the Earth

Tides: The rise and fall of the Earth's water due to the gravitational pull of the moon and the sun

Summary

1. The sun and the eight planets revolving around the sun together form the solar system.
2. The Earth has three main layers known as the crust, the mantle and the core.
3. The sun is a star that is situated at the centre of the solar system and it is a huge ball of burning gases.
4. The moon is the natural satellite of the Earth.
5. Lunar eclipse occurs when the Earth comes between the sun and the moon.
6. Solar eclipse occurs when the moon passes between the sun and the Earth.
7. Spring tides occur when the sun, the moon and the Earth are in a straight line on new moon days.
8. Neap tides occur when the sun is at a right angle to the moon.

EXERCISE

A Choose the correct option for each of the following.

1. The coldest planet of the solar system is
   (a) Neptune (b) Uranus (c) Mars (d) Earth

2. Major constituents of the sun are
   (a) hydrogen and helium (b) neon and oxygen
   (c) helium and oxygen (d) hydrogen and oxygen

3. Solid rocks are found in the outer region of the
   (a) core (b) corona (c) crater (d) mantle
4. Solar eclipse occurs when
   (a) the moon comes between the Earth and the sun
   (b) the Earth comes between the sun and the moon
   (c) the sun comes between the moon and the Earth
   (d) none of the above
5. Lunar eclipse occurs when
   (a) the moon comes between the Earth and the sun
   (b) the Earth comes between the sun and the moon
   (c) the sun comes between the moon and the Earth
   (d) none of the above

B State whether the following statements are true (T) or false (F).

1. The inner core of the Earth is a solid sphere. ☐
2. In the core of the sun, hydrogen gets converted into helium. ☐
3. The moon produces its own light. ☐
4. The moon has a lot of craters on its surface. ☐
5. Total solar eclipse is visible from all parts of the Earth. ☐

C Fill in the blanks using the correct words.

1. The outermost layer of the Earth is called the __________. (crust/core)
2. During volcanic eruptions, molten magma comes out from the __________ of the Earth. (crust/mantle)
3. The moon is the __________ satellite of the Earth. (natural/artificial)
4. The __________ has no atmosphere. (Earth/moon)
5. Spring tides are seen on __________ days. (full moon/new moon)

D Answer the following in one word.

1. Name the fixed path on which planets revolve around the sun. __________
2. What is the temperature on the surface of the sun? __________
3. On which layer of the Earth do oceans lie? __________
4. Name one element present in the core of the Earth. __________
5. What is the ring-shaped eclipse called? __________

E **Answer the following questions briefly.**

1. What is the solar system?
2. What are artificial satellites?
3. What is an eclipse? Name the two eclipses that are visible to us.
4. What is the difference between a total solar eclipse and a partial solar eclipse?
5. What is an annular eclipse?
6. What are tides? How are they formed?

F **Answer the following questions.**

1. Write a note on the features of the sun.
2. Explain the different layers of the Earth.
3. Explain the features of the surface of the moon.
4. How does a lunar eclipse occur?
5. How does a solar eclipse occur?

G **HOTS (Higher Order Thinking Skills)**

1. What is the special feature of the Earth compared to other planets in the solar system? (Hint: Think about the things that are found only on the Earth.)
2. A lunar eclipse occurs only on a full moon day. Why?
   (Hint: What is the position of the sun, the moon and the Earth during a lunar eclipse?)
3. Can we hear someone speak on the moon?
   (Hint: Sound waves cannot travel without a medium.)

H **Label the different layers in the internal structure of the Earth.**

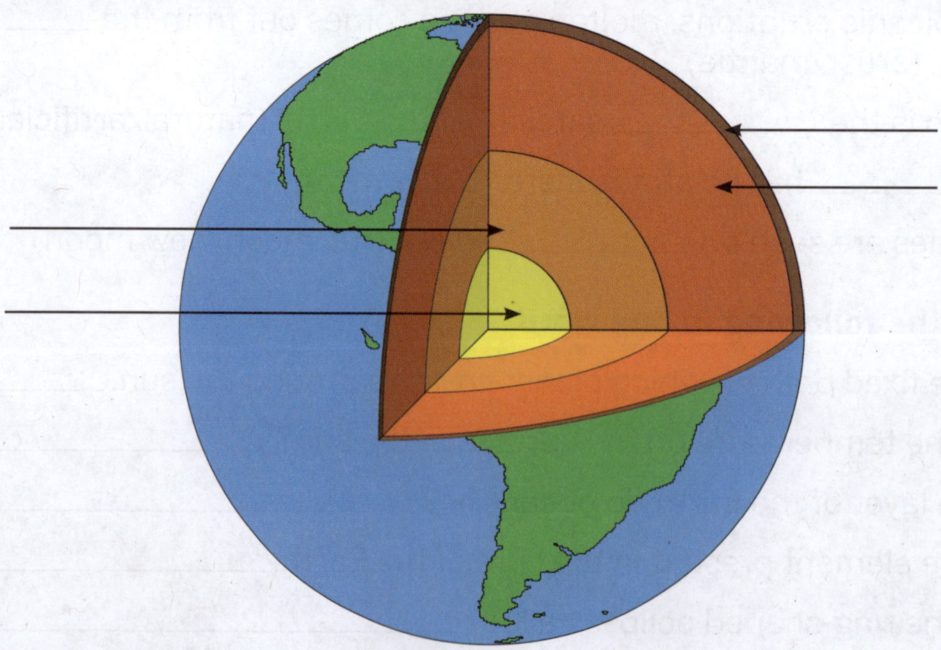

I Project

1. Using a torch, a big ball, a small ball and a screen, we can demonstrate the lunar eclipse. Place the torch and the screen on a table. Light the torch and place a big ball between the torch and the screen.

   Now, place the small ball at point A as shown in the diagram. You can see the shadows of both the balls on the screen. Move the small ball along the line from A to B. You cannot see the shadow of the small ball when it enters in the shadow of the big ball, and again as you move the ball further, the shadow appears.

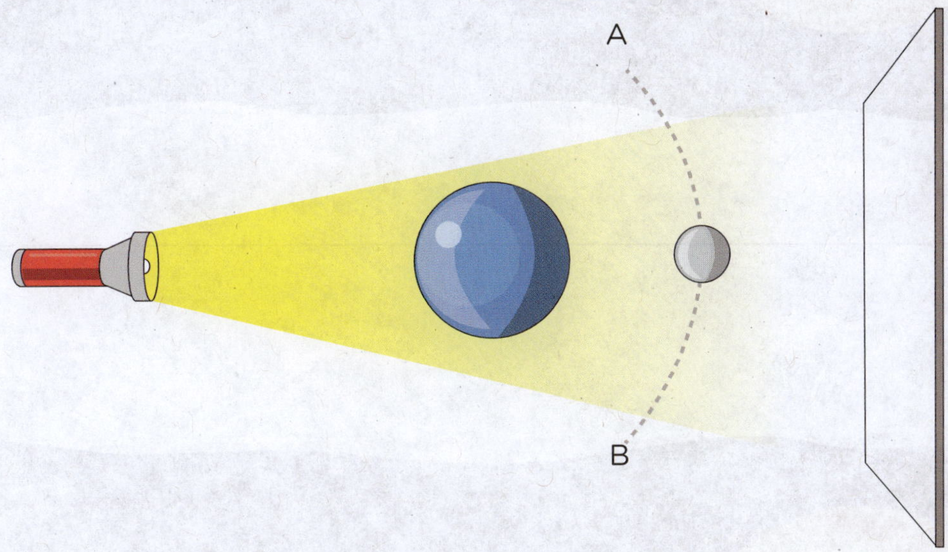

2. Collect interesting facts about the surface of the moon. Also, collect information about the various exploration missions to the moon. Prepare a chart and present the information you have collected to the class.

J Life Skills

The Earth is the only planet on which life is known to exist. It is at the right distance from the sun and has an atmosphere that protects it from the harmful rays of the sun. It also has green plants that produce oxygen during photosynthesis. Oxygen is essential for living things to survive.

Suppose the astronauts who land on the moon decide to stay there forever. Would they be able to survive there? Give reasons for your answer.

WORKSHEET

Label the diagrams given below and name the phenomena they represent.

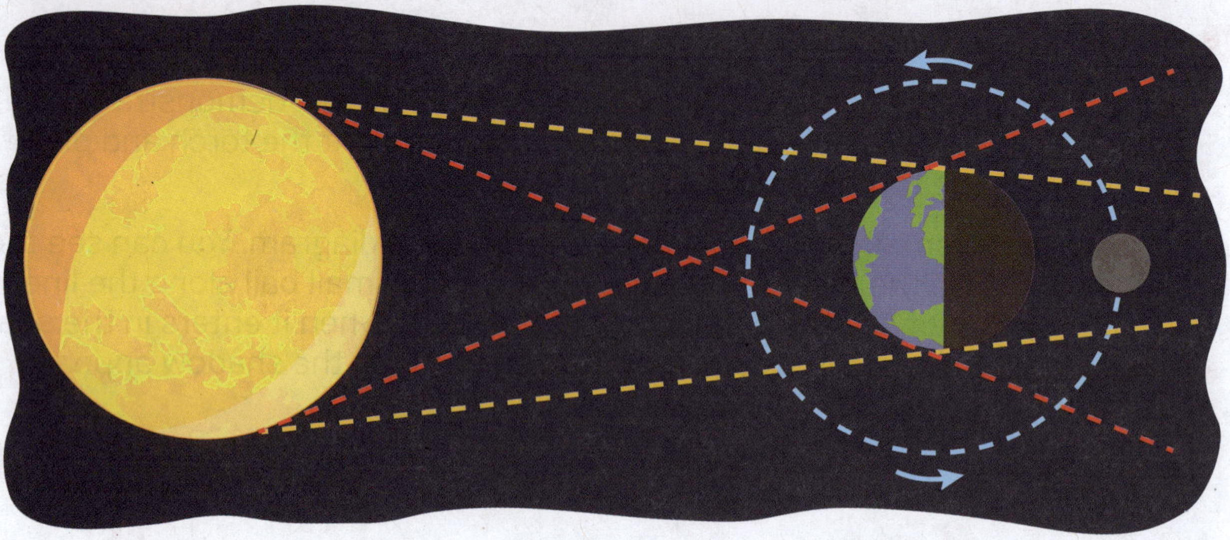

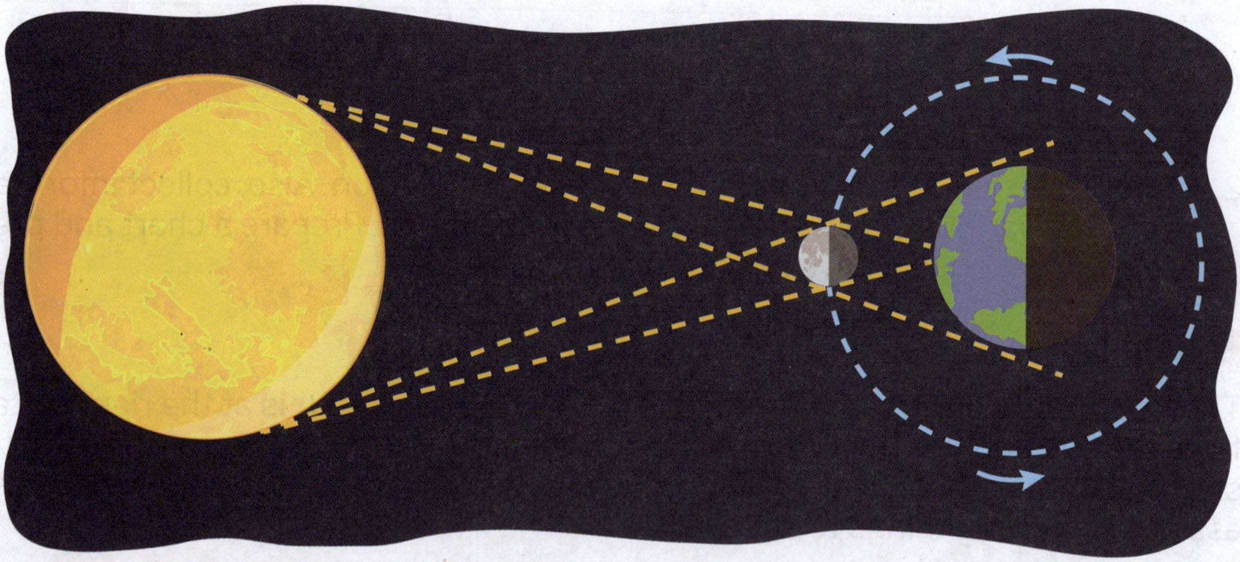

8 Our Environment

Learning Checklist

- Pollution of air, water, land and noise
- Saving the environment

Let's Get Started

Human beings have made their environment dirty in many ways. Look at the pictures given below and tick the ones that make the environment dirty.

ENVIRONMENT

The things around us form our environment.

Our environment contains both living things and non-living things. All living things, including us, depend on the non-living things. For example, we need air to breathe and water to drink. We get these from the environment around us.

The food we eat also comes from plants and animals. We get raw materials like soil, sand, stones, etc. from the Earth to build our homes and buildings.

A natural environment

POLLUTION

The addition of harmful substances to the environment which makes it dirty is called **pollution**.

The substances that cause pollution are called **pollutants**. Pollution has harmful effects on plants, animals and human beings.

AIR POLLUTION

Air in the cities contains a lot of impurities. Smoke from industries and vehicles is the main source of air pollution. In the villages, many houses still use firewood and coal for cooking. This gives out smoke causing air pollution in the rural areas. Smoke from firecrackers also contributes to air pollution. People who breathe in the polluted air suffer from respiratory diseases like asthma, breathing problems and lung cancer.

Bursting of firecrackers causes air pollution

Burning of wood causes air pollution

Smoke from factories and vehicles causes air pollution

Gases like carbon dioxide and methane trap the heat from the sun and do not allow it to escape from the Earth's atmosphere. This is called the **greenhouse effect**. Over the years, increase in the amount of carbon dioxide in the atmosphere due to air pollution has lead to a gradual rise in the Earth's temperature. This is called **global warming**. Global warming can cause the ice at the polar regions to melt, leading to a rise in the water level of the oceans, which results in floods.

WATER POLLUTION

The main sources of water pollution are the wastes from industries, and the sewage pipelines that come from homes and other buildings. In the villages, the lakes or rivers are polluted due to the various activities that are done in or near the water source. Villagers depend on the lake or river for washing clothes and bathing cattle. The dirt and soap from all this washing goes into the water and causes pollution. The solid wastes dumped in the water also causes water pollution.

Sewage let into the water source

The seas and oceans have a threat from oil spillage from ships that transport oil. The oil does not mix with the water, and harms the lives of the plants and animals living in these water bodies.

Various causes of water pollution

Pollution of water makes it unfit for human use. Sewage contains harmful germs that cause diseases like cholera, diarrhoea and typhoid. If these germs are present in drinking water, they can cause diseases.

Science in Context

The Exxon-Valdez oil spill in Alaska happened on 24 March 1989. A tanker was carrying 55 million US gallons of crude oil. The tanker met with an accident causing the spillage of oil into the seawater. This caused the loss of life of animals like sea otters, seals, bald eagles and seabirds. Fish population in the sea was disturbed and the decrease in fish population caused a loss to the people working in the fishing industry.

LAND POLLUTION

Dumping wastes like plastic bags, glass and metal containers causes land pollution. Overuse of chemicals such as fertilizers, insecticides and pesticides also causes land pollution.

Soil is held together by the roots of trees. Deforestation or cutting of trees makes the soil loose. This loose soil is easily carried away by the blowing wind. The rainwater also washes away the loose soil. The process of removal of the top layer of soil by wind and rainwater is called **soil erosion**.

Garbage dumped in the soil causing land pollution

Soil erosion

Revise and recall

Fill in the blanks with the correct words.

1. Smoke from automobiles causes ___________ pollution. (land/air)
2. The wastes from industries and the sewage pipelines are the sources of ________________ pollution. (water/air)
3. Dumping wastes like plastic bags, glass and metal containers causes ________________ pollution. (air/land)
4. The roots of trees hold the ________________ together. (soil/water)

NOISE POLLUTION

Noise pollution is caused by the loud sounds that disturb the normal activities of human beings and animals.

Honking of vehicles, sound from the loud speakers, television, etc. cause noise pollution.

Noise from firecrackers, and construction of buildings and roads also cause noise pollution.

Noise pollution has many adverse health effects such as hypertension, hearing impairment, heart disease and sleep disturbance.

Noise pollution in a city

SAVING THE ENVIRONMENT

We should try to minimize pollution to save our environment. Some ways of reducing pollution are given below:

1. Reducing air pollution by using the vehicles only when needed, and walking or cycling to short distances.
2. The smoke and gases produced from the factories should be processed and made safe before letting out through chimneys.
3. Planting more trees to reduce air pollution to get pure air to breathe and to prevent soil erosion.
4. Treatment of waste from factories and sewage from homes before being let into water bodies.
5. Proper management and disposal of garbage.
6. Using environment-friendly manures and natural fertilizers in fields.
7. Using cloth bags instead of polythene bags and minimizing the use of plastics.
8. Construction of factories far away from the main city.
9. Use of solar energy should be encouraged as it does not cause pollution.

Let's Say Together!

Environment	Pollution	Smoke
Global warming	Sewage	Erosion

Keywords

Pollution: The addition of harmful substances to the environment which makes it dirty
Pollutants: The substances that cause pollution
Greenhouse effect: The trapping of heat by the gases in the atmosphere that leads to increase in the Earth's temperature
Global warming: A gradual rise in the Earth's temperature due to greenhouse effect

Summary

1. All the living and non-living things in our surroundings form our environment.
2. Smoke and gases from industries and automobiles, firecrackers, burning of firewood, etc. cause air pollution.
3. The main sources of water pollution are sewage and waste from the industries.
4. Disposal of garbage or waste and overuse of chemicals like insecticides, fertilizers and pesticides are the main causes of land pollution.
5. Building construction, honking of vehicles, sound from the loudspeakers, etc. cause noise pollution.
6. Planting more trees and safeguarding the forests reduce pollution.
7. Stopping water pollution by proper treatment and safe waste disposal can reduce water and land pollution.

EXERCISE

A Choose the correct option for each of the following.

1. Which of the following is a cause of air pollution?
   (a) Trees (b) Smoke (c) House (d) All of these
2. Which of the following is not a cause of soil erosion?
   (a) Wind (b) Rain (d) Trees (d) None of these
3. Air pollution can cause
   (a) asthma (b) obesity
   (c) hearing impairment (d) cholera

4. Which of these causes land pollution?

   (a) Loudspeakers (b) Polythene bags

   (c) Smoke (d) Sewage

5. Honking of vehicles and loud noise from television cause ________ pollution.

   (a) land (b) water (c) air (d) noise

B Fill in the blanks using the correct words.

1. The substances that cause pollution are called ____________. (pollutants/smoke)
2. Global warming is caused due to increase in the amount of ____________. (carbon dioxide/oxygen)
3. The main cause of soil erosion is ____________. (deforestation/afforestation)
4. Washing clothes and bathing animals near a water source cause ____________ pollution. (water/air)
5. Noise pollution causes ________________________. (sleep disturbance/asthma)

C State whether the following statements are true (T) or false (F).

1. Burning firecrackers causes air pollution. ☐
2. Global warming can cause melting of ice in the polar regions. ☐
3. Oil spillage from ships cause water pollution. ☐
4. Sewage contains germs that cause cholera, diarrhoea, etc. ☐
5. Pollution of air, water, noise and land is beneficial for us. ☐

D Give two examples for each of the following.

1. Kinds of pollution ____________ ____________
2. Activities that cause water pollution ____________ ____________
3. Causes of soil erosion ____________ ____________
4. Wastes that cause land pollution ____________ ____________
5. Health effects of noise pollution ____________ ____________

E Answer the following questions briefly.

1. What is pollution?
2. What are pollutants?
3. Write the harmful effects of air pollution.
4. What is global warming?
5. What is soil erosion?

F Answer the following questions.

1. Write a note on air pollution.
2. What are the causes of water pollution in cities and villages?
3. Describe the sources of noise pollution.
4. Write any five ways to save our environment.

G HOTS (Higher Order Thinking Skills)

1. What does the sign given alongside signify? Where is this sign usually found? (Hint: Noise pollution)
2. If firecrackers are burst, they cause two types of pollution. What are they? (Hint: Firecrackers make loud sound and give out smoke.)
3. We must protect and conserve our environment. Justify the statement. (Hint: Effects of pollution)

H Tick the things that can be used to reduce pollution.

 ☐

 ☐

 ☐

 ☐

 ☐

I Project

Make posters on the four types of pollution you have studied in this chapter and ways to prevent them. Write catchy slogans to create awareness.

J Life Skills

1. Where would you throw the empty plastic juice bottles?

   (a) Dustbin (b) River water (c) Soil

2. When should you use the bell on your cycle?

   (a) When you want to alert people crossing the road

   (b) To have fun (c) Near the hospital

3. To reduce air pollution and to get fresh air, you should

   (a) plant trees (b) burst firecrackers (c) use cars for travel

Identify the type of pollution shown in the following pictures and write in the space provided.

______________________ ______________________

______________________ ______________________

9 Natural Disasters

Learning Checklist

- Earthquake
- Volcanoes
- Tidal waves and tsunami
- Floods
- Drought

Let's Get Started

Look at the pictures shown below. Can you identify the given phenomena? Choose the words from the help box.

Earthquake	Drought	Thunderstorm	Volcanic eruption

____

____

____

____

A **disaster** is defined as an unexpected event resulting in great damage, loss and destruction to life and property. Natural events occurring on the Earth that cause great damage or loss of life are called **natural disasters**.

In this chapter, we will learn more about natural disasters.

EARTHQUAKE

The sudden movement of a part of the Earth's surface is called an **earthquake**. These movements are caused by shifting of layers of rocks deep under the Earth's crust. The actual point deep inside the Earth where an earthquake begins is called the **focus** and the point on the Earth's surface directly above the focus is called the **epicentre**. The maximum effect of the earthquake is felt at the epicentre and the effect reduces as we move away from the epicentre.

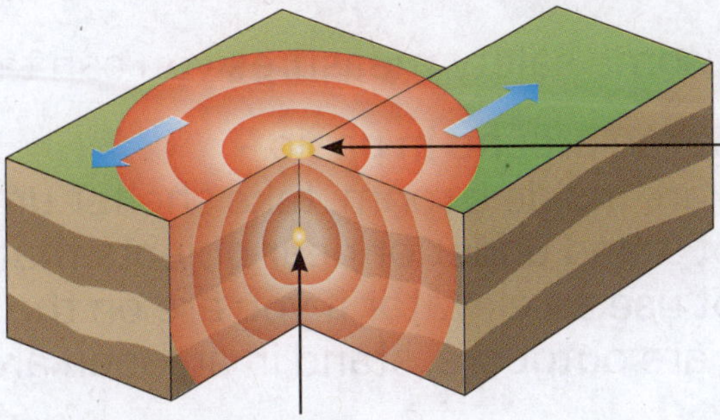

After-effects of Earthquakes

1. Destruction of buildings, trees, roads and vehicles
2. Huge loss of life (humans and animals)
3. Spread of diseases

Destruction caused by earthquakes

Measuring Earthquakes and Their Effects

The effects of earthquakes depend on the force with which they occur. Some earthquakes are so weak that they just shake materials kept on a shelf, while some are so strong that they cause buildings to fall down and lead to deaths of numerous people.

The intensity and duration of an earthquake can be measured using an instrument called the **seismograph**. **Richter scale** is a standard scale used to find out the magnitude[1] of earthquakes on a scale of 1 to 10. Earthquakes measuring 1 to 2 on the scale will hardly be felt. Earthquakes measuring 3, 4 and 5 on the scale can cause damage to property. Earthquakes measuring 6 and above on the scale can result in widespread damage to property and also cause death.

Word Web

[1]**Magnitude**: relative size or extent

Modern seismograph

Know the Fact!

Zhang Heng of China's Han dynasty was the first person to build an instrument to detect earthquakes.

Safety Measures During an Earthquake

Here are some safety measures you can take to be safe during an earthquake:

- Do not run around in panic.
- If you are inside a room, take shelter under a table or a bed.
- Do not stand under or near objects that can fall and hurt you.
- Do not use an elevator if you are on the upper floors. Always use the staircase.
- If you are outdoors, stand in an open area away from any buildings and trees.

VOLCANOES

The core of the Earth is made up of molten rock called **magma**. A crack or vent on the Earth's surface that allows hot magma from below the surface to come out is called a **volcano**.

A **vent** is the region where the Earth's crust is weak and develops a crack. Once hot magma comes out of the Earth's surface, it is called **lava**. After a volcanic eruption occurs, a crater is formed at the region of vent.

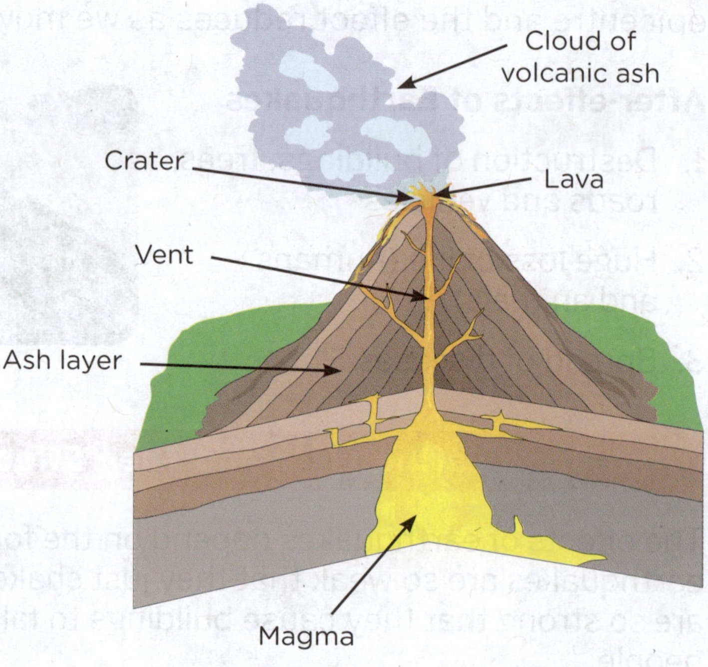

A volcano

Volcanoes are of three types depending on how frequently they erupt: active, dormant and extinct.

Active volcanoes: These are considered to be highly dangerous. They may erupt at any time or have erupted in recent years. Some of the most active volcanoes are Mount St. Helens in the United States, Mount Mayon in the Philippines and Mount Lassen in the United States.

Dormant volcanoes: Volcanoes that have not erupted for a long period but may erupt in the future are called dormant volcanoes. Some of the dormant volcanoes are Mauna Kea in Hawaii, Mount Edziza in Canada and Narcondam in Andaman Islands of India.

Extinct volcanoes: Volcanoes that are not expected to erupt in the future are called extinct volcanoes. Some of the extinct volcanoes are Mount Ashitaka in Japan, Shiprock in the United States and Zuidwal in the Netherlands.

After-effects of Volcanic Eruptions

1. Destruction of farmlands and roads, damage to property and loss of life.
2. Volcanic dust blocks sunlight for a long period, which leads to change in the weather pattern.
3. Spread of diseases

Know the Fact

A region in the Pacific ocean that consists of 452 volcanoes is called **the ring of fire**.

Advantages of Volcanic Eruptions

1. After a volcanic eruption, the soil becomes very fertile and good for farming.
2. Some of the volcanic locations can be used to generate geothermal energy[1].

Word Web

[1]**Geothermal energy**: energy obtained from the Earth

TIDAL WAVES AND TSUNAMIS

Tides are the cyclic rise and fall of sea levels caused due to the gravitational pull of the sun and the moon. At times the waves on the sea are high and some times they are low. Tides reach the maximum height when the sun, the moon and the Earth are in a straight line.

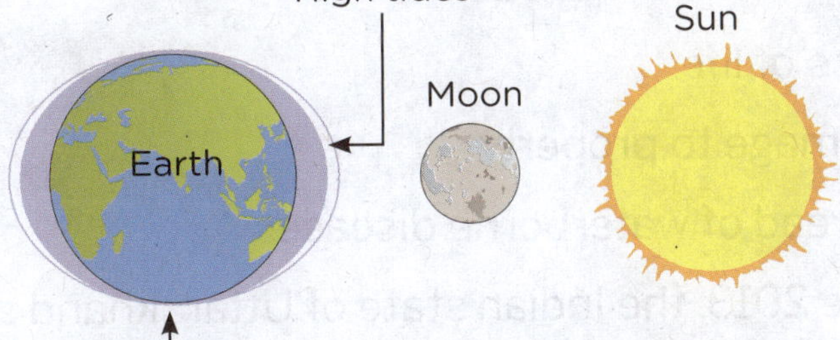

Rise of large amount of water in the form of huge waves from oceans due to underwater earthquakes, volcanic eruptions, etc. is called **tsunami**. When there is a disturbance in the layers of rocks in the Earth's crust under water bodies, a large amount of energy is released, which leads to the formation of huge waves.

Tsunami waves

Large tsunami waves lead to the loss of life and property. It is advisable to stay at high altitude regions during a tsunami threat.

On 26 December 2004, a tsunami occurred in the Indian Ocean, causing a major loss of life and property in the coastal regions of Tamil Nadu and Kerala. The Andaman and Nicobar islands were among the worst hit areas in India.

FLOODS

Approximately 75% of the Earth's surface is covered by water. When water bodies like the oceans, rivers, lakes and ponds have excess water, they overflow leading to floods.

Flood

Floods occur due to the following natural factors:

- Heavy rainfall
- Low-pressure region formed in the ocean
- Excess melting of snow in the mountains and glaciers

Floods can be prevented by taking the following measures:

- Constructing proper streams or channels for the excess water to flow through
- Planting enough trees to slow down the force with which water flows
- Constructing proper embankments[1] that control the level of water flowing into the rivers

Word Web

[1]**Embankment**: a wall made of mud and stone to keep water back

After-effects of Floods

- Loss of life
- Damage to property
- Spread of waterborne diseases

In June 2013, the Indian state of Uttarakhand and nearby places received heavy rainfall, leading to devastating floods and landslides. It caused heavy damage to life and property.

Safety Measures During Floods

Some safety measures you can take to be safe during a flood are given below:

- Do not panic. Always listen to flood warnings given on radios and news channels.
- Leave the area where the flood water is bound to flow.
- Shift to higher locations where the flood water cannot reach.
- Keep medicines, food supplies and clean drinking water in plenty.
- Follow instructions given by rescue team workers.

DROUGHT

Drought

Drought is a condition or state of severe shortage of water supply (both surface and underground). Some droughts are of shorter duration, whereas some continue for years.

Droughts occur due to lack of or insufficient rainfall along with excessive usage of existing water resources.

Effects of drought

Some effects of droughts are as follows:

- Destruction of crops
- Malnutrition and death of humans and animals
- Mass movement of people from one place to another

The effects of drought can be controlled by taking the following steps:

- Making judicious[1] usage of available water resources
- Adapting to rainwater harvesting
- Recycling of water
- Cultivating crops that require less water

Word Web

[1]**Judicious**: careful and sensible

Major drought-prone regions in India are Southern and Eastern Maharashtra, Northern Karnataka, Andhra Pradesh, Odisha, Gujarat and Rajasthan.

In 2012, the Indian state of Karnataka faced severe drought.

Science in Context

Drip irrigation is a modern irrigation technique that saves water. It is done using narrow tubes that allow very small quantity of water to drip into the roots of plants.

Keywords

Earthquake: The sudden movement of a part of the Earth's surface

Volcano: A crack or vent on the Earth's surface that allows hot magma from below the surface to come out

Tide: The cyclic rise and fall of sea levels caused by the gravitational pull of the sun and the moon

Tsunami: Rise of large amount of water in the form of huge waves from oceans due to underwater earthquakes, volcanic eruptions, etc.

Drought: A state of severe shortage of water supply (both surface and underground)

Flood: A natural disaster caused by the overflow of oceans, rivers, etc., due to excess water

Summary

1. Natural events occurring on the Earth that cause great damage to life and property are called natural disasters.
2. The actual point deep inside the Earth where an earthquake begins is called the focus and the point on the Earth's surface directly above the focus is called the epicentre.
3. Intensity and duration of an earthquake can be measured using an instrument called seismograph.
4. Volcanoes are of three types: active, dormant and extinct.
5. Tides reach the maximum height when the sun, the moon and the Earth are in a straight line.
6. Disturbance in the layers of rocks in the Earth's crust under water bodies leads to generation of huge waves called tsunami.
7. When the water bodies like oceans, rivers, lakes and ponds have excess water, they overflow, leading to floods.
8. Drought occurs due to lack of or insufficient rainfall along with excessive usage of existing water resources.

EXERCISE

A Choose the correct option for each of the following.

1. Which of the following natural disasters is caused by disturbances in the Earth's crust?

   (a) Earthquake (b) Flood (c) Tide (d) Drought

2. The actual point deep in the Earth's crust where an earthquake begins is called

   (a) Mantle (b) Core (c) Focus (d) Epicentre

3. Which volcanoes are capable of erupting?

   (a) Dormant and active (b) Active and extinct

   (c) Extinct and dormant (d) Active, dormant and extinct

4. Which of the following natural disasters is caused by shortage of water?

   (a) Drought (b) Tsunami (c) Earthquake (d) Volcanic eruption

5. Heavy rainfall can cause

   (a) Earthquake (b) Drought (c) Volcanic eruption (d) Flood

B State whether the following statements are true (T) or false (F).

1. Tides are formed due to the gravitational pull of the sun and the moon. ☐
2. Drought occurs when water bodies overflow due to excess water. ☐
3. The effect of an earthquake is felt more at the epicentre. ☐
4. Tides and tsunami are one and the same. ☐
5. When magma comes out of the Earth's surface, it is called lava. ☐

C Fill in the blanks using the correct words.

1. The innermost layer of the Earth is called __________. (core/mantle)
2. The magnitude of an earthquake is measured using __________. (seismograph/ Richter scale)
3. Volcanoes that erupt frequently are called __________ volcanoes. (active/extinct)
4. Floods occur due to __________ water. (excess/lack of)
5. Earthquake measuring __________ on the Richter scale is considered to be highly dangerous. (3/8)

D Name the following.

1. The actual point inside the Earth where an earthquake begins __________
2. Huge waves formed due to an earthquake occurring underwater __________
3. The instrument used to measure the intensity and duration of earthquakes __________
4. The region where the Earth's crust is weak and develops a crack __________
5. The natural disaster that occurs due the lack of rainfall __________

E Answer the following questions briefly.

1. What is an earthquake?
2. What is an epicentre?
3. What is a dormant volcano?
4. What causes tides?
5. What is a tsunami?

F Answer the following questions.

1. What are the after-effects of an earthquake? Give three safety measures to be followed during earthquakes.
2. What is a volcano? Explain different types of volcanoes with one example each.
3. Write two after-effects and two advantages of volcanic eruption.

4. When do floods occur? What are the after-effects of flood?
5. Give the measures that can be taken to prevent or reduce the effects of drought.

G HOTS (Higher Order thinking Skills)

1. Amit is on the first floor of a building. If an earthquake occurs suddenly, which method is better for him to come out of the building: using the stairs or the elevator? Why? (Hint: Recall the safety measures during an earthquake.)
2. Can an earthquake cause a tsunami? Explain. (Hint: What is a tsunami?)

H Label the parts of the volcano shown below.

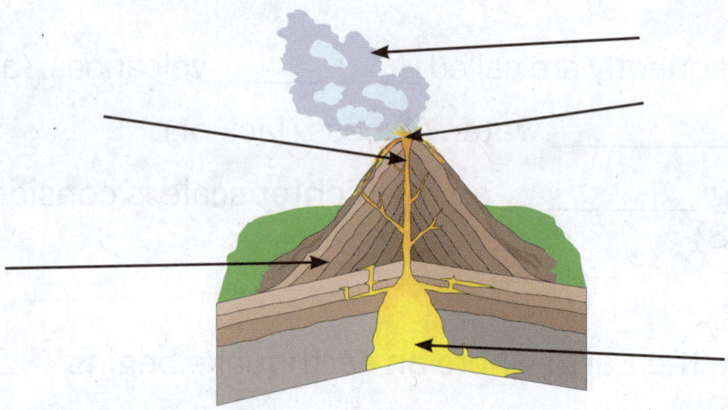

I Project

Search the Internet or newspapers with the help of an elder and make a report on natural disasters. Cover the following points:

(a) Name of the disaster
(b) The date on which it occured
(c) Places affected by the disaster
(d) Details of damage or loss of life caused, if available

J Life Skill

Suppose you are inside your house and you experience a slight shake. Suddenly you realize that it is an earthquake. Read the following points to know what you should do to save yourself.

- Drop down; take shelter under a bed or a table and hold on.
- Stay indoors until the shaking stops and you are sure it is safe to exit.
- Stay away from bookcases or furniture that can fall on you.
- Stay away from windows. In a high-rise building, expect the fire alarms and sprinklers to go off during a quake.
- If you are on bed, hold on and stay there, protecting your head with a pillow.
- If you are outdoors, find a clear spot away from buildings, trees and power lines. Drop down to the ground.

Identify the natural disasters shown below. Write a few sentences on the after-effects of each.

WORKSHEET

10 Rocks and Minerals

Learning Checklist

- Types of rocks
- Minerals
- Uses of rocks and minerals

Let's Get Started

Look at the pictures given below and fill in the blanks with the given words.

Chalk Ruby Granite

__________ is used to make kitchen slabs.

__________ is used to write on blackboards.

__________ is used in making jewellery.

Rocks and minerals are of different shapes, sizes and colours. They are found everywhere on the Earth. Let us learn more about them.

TYPES OF ROCKS

Rocks can be divided into three groups based on how they are formed: igneous, sedimentary and metamorphic.

Igneous Rocks

We have learned in Chapter 9 that the core of the Earth is made of molten rock called **magma**. When magma comes out to the surface of the Earth through volcanic eruptions, it is called **lava**.

A volcanic eruption

The lava cools down and hardens to form igneous rocks. Some of the minerals present in igneous rocks are mica, feldspar, magnesium and quartz. Igneous rocks derive their name from the Latin word *ignis* which means 'fiery'.

Granite, pumice and obsidian are examples of igneous rocks.

Granite is formed due to the slow cooling of lava. The three main minerals found in granite are mica, feldspar and quartz. The colour of granite varies from grey to pink based on the amount of feldspar present. Granite is used to make buildings, interiors of kitchens, floorings, etc.

Pumice is formed by the rapid cooling of lava. It is creamish or grey in colour. The release of gases from the lava forms the pores in the pumice rock.

Unlike other rocks, pumice is lightweight. It is used in making lightweight building materials, footscrubs, etc.

Science in Context

Pieces of pumice rocks can float on water because the air-pores present in them make them less dense than water.

Obsidian is also formed due to the rapid cooling of lava. It is a dark or black-coloured rock. It is smooth and glassy; hence it is also called **volcanic glass**. It has very sharp edges. It is used in making jewellery.

Granite Pumice Obsidian

Igneous rocks

Sedimentary Rocks

Sedimentary rocks are formed due to the deposition of small pieces of rock, soil, and plant and animal matter over thousands of years by the action of wind, rainwater and flowing river water.

Rivers flowing down a mountain wear away rocks. Wind and rainwater also wear away rocks. These small pieces of rocks are deposited at some other place. Layer over layer is deposited over a long period of time. This repeated deposition hardens the lower layers, leading to the formation of sedimentary rocks.

Conglomerate, sandstone, shale and limestone are examples of sedimentary rocks.

Conglomerate contains gravel, round pebbles and boulders held together by minerals. It has an uneven texture and is used as a building material.

Limestone contains the mineral calcium and tiny animal shells. Its grains are soft and fine. It is used to make lime, cement, paint, etc.

Sandstone is formed by the deposition of sand particles held together by minerals like silica. Famous Indian monuments like the Red Fort and the Qutub Minar are made of sandstone.

Shale contains solidified clay and mud piled up in layers. It is used to make construction materials like bricks, tiles and cement.

Conglomerate

Limestone

Sandstone

Shale

Sedimentary rocks

Metamorphic Rocks

Metamorphism means a gradual change in form. Igneous rocks and sedimentary rocks change into metamorphic rocks over time due to high pressure and temperature.

Slate, marble, quartzite and gneiss are examples of metamorphic rocks.

Slate is formed from the sedimentary rock shale. Slate mainly contains clay, mica and other minerals. It can be broken into thin layers. It is used to make slates for children, blackboards for classrooms and cement.

Marble is formed from the sedimentary rock limestone. It has been widely used for construction from ancient times. One of the best examples of this is the Taj Mahal in Agra, which is made of white marble. Marble is also used in making flooring material and statues.

Quartzite is formed from the sedimentary rock sandstone. It mainly contains iron, quartz and silica. These are hard rocks and are not easily broken. It is used in making floorings and stairs, and in the glass and ceramic industries.

Gneiss is formed from the igneous rock granite. It has light and dark bands of varying length and thickness. It is used as a building material.

Slate

Marble

Quartzite

Gneiss

Metamorphic rocks

Revise and recall

Match the following.

Column A	**Column B**
(a) Taj Mahal is made of	(1) Obsidian
(b) Molten rock inside the Earth's core	(2) Sandstone
(c) Lightweight rock with holes	(3) Marble
(d) The rock used in building Red Fort	(4) Magma
(e) Volcanic glass	(5) Pumice

MINERALS

Rocks are made up of substances called minerals. Minerals are formed inside the Earth.

There are two types of minerals: metallic and non-metallic.

Metallic Minerals: These minerals yield metals. Metal-yielding minerals are also called **ores**. A few examples of ores are bauxite, pyrite and hematite. These ores can be processed to obtain aluminium, copper and iron, respectively.

Metals are used in making machines, cars, ships, wires, etc. Gold, silver and platinum are precious metals that are used to make jewellery and ornaments.

Gold

Iron bars used in construction

Copper

Metals obtained from minerals

Non-metallic Minerals: Some important non-metallic minerals are coal, salt, gypsum, mica and feldspar.

Coal is used in the thermal power generation, for cooking and in other industries where a large amount of heat is required.

Coal

Emerald (a gemstone)

Non-metallic minerals

Gemstones are also obtained from minerals. Pieces of these minerals are cut and polished for using them to make jewellery. These gemstones are found in very attractive colours like red, green and yellow.

A few examples of gemstones are diamond, ruby, sapphire and emerald. These are also known as precious stones.

Science in Context
Diamond is the hardest mineral and can withstand very high temperatures.

USES OF ROCKS AND MINERALS

Rocks and minerals are being used by human beings from ancient times. Even to this day, we use products obtained from the rocks and minerals.

Following are some of the most common uses of rocks and minerals.

Construction: We use rocks and stones for building houses, offices, etc. Stones and jelly stones are used all through the building process. Even the cement used in construction of buildings is derived from minerals.
Tar is used to make roads, and marbles are used for flooring. Iron and steel are used in making pillars for the buildings.

Metals: Utensils are the basic requirements of a kitchen in a home. These are made from iron, copper, aluminium and steel. All these are made or obtained from the mineral ores.

Metals like gold, silver and platinum are used in making ornaments, decorative items and jewellery.

Gemstones: Gemstones like diamond, ruby, sapphire and emerald are used to make jewellery.

Fuels: Coal is used as a fuel to keep homes warm in colder regions and for cooking. Uranium is a rare mineral. It is used in nuclear reactors to produce large amount of energy.

Fertilizers: Minerals like sulphates and nitrates are constituents of fertilizers. Fertilizers are added to the soil to increase the fertility of soil, thereby increasing the yield of crops.

Food: We use rock salt (*kala namak*) in food. It is obtained from the rocks in the Earth.

Many other minerals like calcium, iron and potassium are required for the growth and development of our body. We get these from the vegetables and fruits we eat. Plants obtain these minerals from the soil.

The following table lists some minerals and their use in our daily lives.

Minerals	Used in
Quartz	Watches
Talc	Talcum or face powder
Graphite	Pencils
Diamond	Jewellery
Copper	Electrical wires
Mica	Welding rods, electronic goods

Let's Say Together!

Igneous
Magma
Sedimentary
Ores
Metamorphic
Minerals

Keywords

Crust: The outer surface of the Earth on which we live

Magma: Molten rock present inside the Earth

Lava: Magma flowing on the surface of the Earth after a volcanic eruption

Igneous rocks: These are the rocks formed as a result of cooling and solidification of molten rock

Sedimentary rocks: These are the rocks that are formed due to deposition of small pieces of rock, soil, and plant and animal matter carried by the wind or flowing water

Metamorphic rocks: Igneous rocks or sedimentary rocks undergo changes over a period of time to form these rocks

Ore: Metal-yielding minerals

Summary

1. There are three kinds of rocks: igneous, sedimentary and metamorphic.
2. Igneous rocks are formed due to the cooling and solidification of molten magma.
3. Sedimentary rocks are formed due to the formation of sediments due to wind and flowing water.
4. Metamorphic rocks are formed by the gradual changes that take place in the igneous and sedimentary rocks.
5. Rocks are made up of minerals. Minerals are found inside the Earth. There are two types of minerals: metallic and non-metallic.
6. Rocks and minerals are widely used by human beings.

EXERCISE

A **Choose the correct option for each of the following.**

1. Which of the following is an igneous rock?
   (a) Obsidian (b) Sandstone (c) Limestone (d) Slate
2. The sedimentary rock that contains solidified mud and clay piled up in layers
   (a) Limestone (b) Sandstone (c) Shale (d) Conglomerate
3. The metamorphic rock formed from sandstone
   (a) Gneiss (b) Marble (c) Quartzite (d) Slate
4. Which of the following is a non-metallic mineral?
   (a) Pyrite (b) Bauxite (c) Hematite (d) Gypsum
5. Which of the following is a gemstone?
   (a) Coal (b) Mica (c) Emerald (d) Feldspar
6. Which of the following minerals is used in welding rods and electrical goods?
   (a) Talc (b) Mica (c) Quartz (d) Diamond

B **Fill in the blanks with the correct words.**

1. The molten rock present inside the Earth is called ____________. (magma/lava)
2. Pumice is a type of ____________ rock. (sedimentary/igneous)
3. Gneiss is formed from the igneous rock ____________. (granite/shale)
4. ____________ is used to make roads. (Tar/Marble)
5. ____________ is used in making jewellery. (Graphite/Diamond)
6. ____________ is used in food. (Talc/Rock salt)

C **State whether the following statements are true (T) or false (F).**

1. Granite is formed from limestone. ☐
2. Coal is a metallic mineral. ☐
3. Slate can be broken into thin layers. ☐
4. Shale is used in making bricks. ☐

5. Pumice is formed by the rapid cooling of lava. ☐
6. Minerals like sulphates and nitrates are the constituents of fertilisers. ☐

D Give one example for each of the following.

1. Igneous rock ____________
2. Sedimentary rock ____________
3. Metamorphic rock ____________
4. Mineral ____________
5. Gemstone ____________

E Match the following.

Column A	Column B
(a) Igneous	(1) Marble
(b) Sedimentary	(2) Feldspar
(c) Metamorphic	(3) Obsidian
(d) Non-metallic mineral	(4) Diamond
(e) Gemstone	(5) Conglomerate

F Answer the following questions in brief.

1. Which types of rocks are formed after a volcanic eruption? Give two examples.
2. Which rock is also called volcanic glass? Give one use of this rock.
3. What are the two things that help in the formation of sedimentary rocks?
4. What are the two constituents of sandstone? Name two monuments made of sandstone.
5. What are metamorphic rocks? Give two examples.
6. What are minerals? Name the two types of minerals.
7. What are gemstones? What are they used for?
8. What are ores? Give two examples.

G Answer the following questions.

1. Explain the formation of igneous rocks.
2. Write a note on the formation and uses of pumice.
3. Explain the formation of sedimentary rocks. Give two examples and explain their features.
4. Explain the features of the metamorphic rocks and their formation.

5. Describe the rock quartzite.
6. Give the various uses of rocks and minerals.

H HOTS (Higher Order Thinking Skills)

Romila says gold and silver are metals, but Lekha says they are minerals. Who is correct? Can both be correct? Why or why not?

I Name the rocks the following monuments are made of.

Red Fort

Taj Mahal

Identify the rocks and write whether they are igneous, sedimentary or metamorphic.

J Project

Collect stones of different colours and identify them with the help of your teacher. Keep them in egg trays and make an exhibit for your school science exhibition.

K Life Skills

All the minerals are exhaustible, that is, the availability of these minerals is limited.

Discuss the following in your class.

- What are the difficulties that we would have to face if the minerals get exhausted in nature?
- How can we help to save the minerals for the future?

Write the name of the rock or mineral that is used for making the following products.

Slate

____

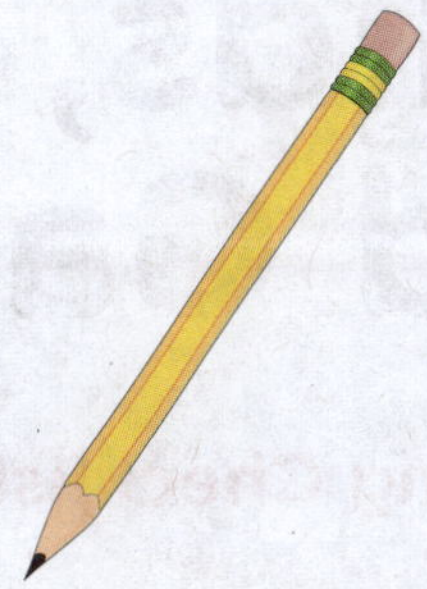

Pencil lead

____

Cement

____

Wristwatch

____

Kitchen slab

____

Talcum powder

____

11 Solids, Liquids and Gases

Learning Checklist

- What is matter made up of?
- Heating and cooling of solids, liquids and gases
- Physical and chemical changes

Let's Get Started

We know that water exists in three forms or states: solid, liquid and gas. Draw a line to match the pictures on the left with those on the right that show the same state of matter.

Everything in the universe is made up of matter. A substance that has mass and occupies space is called **matter**. Matter can exist in three states: solid, liquid and gas. Matter is made up of small particles called **atoms** and **molecules**. Let us learn more about them.

WHAT IS MATTER MADE UP OF?

Atoms are the building blocks of all matter. They are extremely small in size and can be seen only with the help of very powerful microscopes. Though you cannot see them with naked eyes, everything around you is made up of atoms. The water you drink, the air you breathe, the Earth and everything on it is made up of atoms.

Molecules

When two or more atoms combine together, they form a particle called a **molecule**. A molecule can be made up of the same or different kind of atoms. For example, an oxygen molecule is made up of two oxygen atoms (also written as O_2), whereas a water molecule is made up of two hydrogen atoms and one oxygen atom (also written as H_2O).

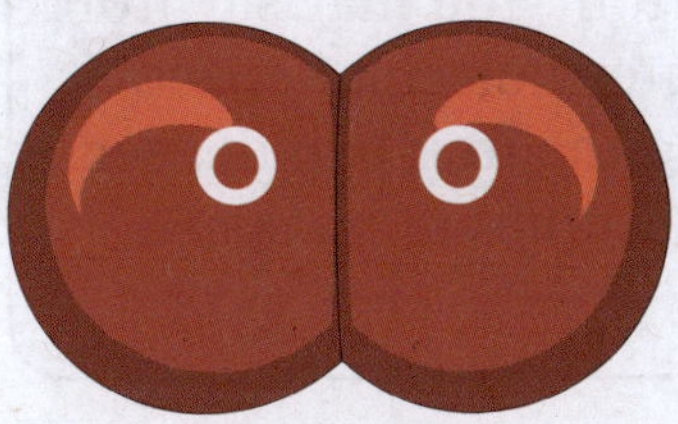

O_2
Oxygen molecule

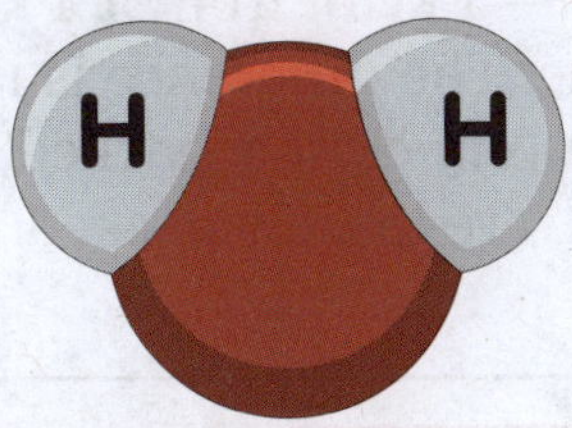

H_2O
Water molecule

Science in Context

Nanotechnology is the branch of science that involves working with atoms and molecules to build devices (such as robots) that are extremely small.

Atoms and molecules are in constant motion, that is, they are always moving. In **solids**, atoms or molecules move but remain in the same position. In other words, they only vibrate without changing their position. That is why solids have a fixed shape. In **liquids**, atoms or molecules move around and glide over each other. That is why liquids flow easily and can be poured into containers. In **gases**, atoms or molecules move freely and are not in contact with each other. That is why gases move to take up the entire space available to them.

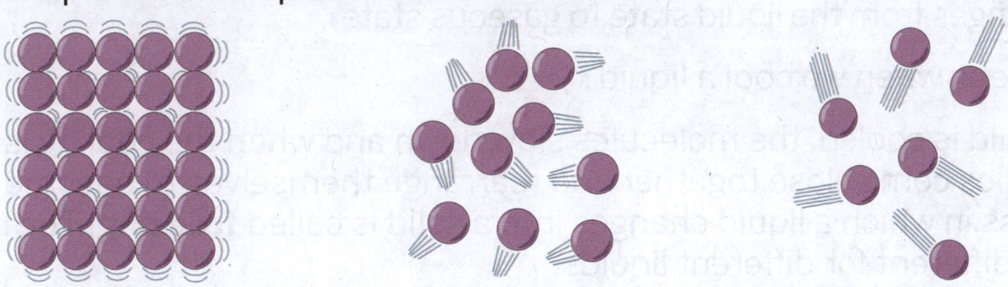
(a) (b) (c)
Movement of particles in (a) solids (b) liquids and (c) gases

Intermolecular Space

The space between two molecules is called their **intermolecular space**.

The molecules are held together by a force of attraction, which is very strong in solids, weaker in liquids and very weak in gases.

A spoon of sugar added to water dissolves in it because the sugar molecules occupy the intermolecular space in water.

Some liquids such as alcohol also dissolve in water. Gases such as oxygen and carbon dioxide dissolve in water. Fish breathe in the oxygen dissolved in water.

Extended Learning

The lightness or heaviness of a substance for a given volume is called **density**. Density depends on how closely the molecules are packed in a given substance. Thus, a stone is denser than a block of wood because the particles in a stone are very closely packed compared to the block of wood.

Science in Context

Have you ever wondered which is heavier — 1 kg of iron or 1 kg of cotton? You may say iron but this is not true as both weigh the same. The difference is in the intermolecular space between the two. The particles of iron are very closely packed, making the iron ball denser. The particles of cotton, on the other hand, are loosely packed. So, it occupies more space but is less dense.

HEATING AND COOLING OF LIQUIDS

What happens when we heat a liquid?

When a liquid is heated, the molecules acquire energy and move faster. When it is heated continuously, the molecules ultimately move so fast that they are no longer in contact with each other and thus, the liquid changes into gas.

This process in which a liquid changes into gas is called **evaporation**. Evaporation causes wet clothes to dry and the water in them to disappear (that is, water in the clothes changes from the liquid state to gaseous state).

What happens when we cool a liquid?

When a liquid is cooled, the molecules slow down and when the cooling is continued, the molecules come close together and rearrange themselves to become a solid. This process in which a liquid changes into a solid is called **freezing**. The rate of freezing is different for different liquids.

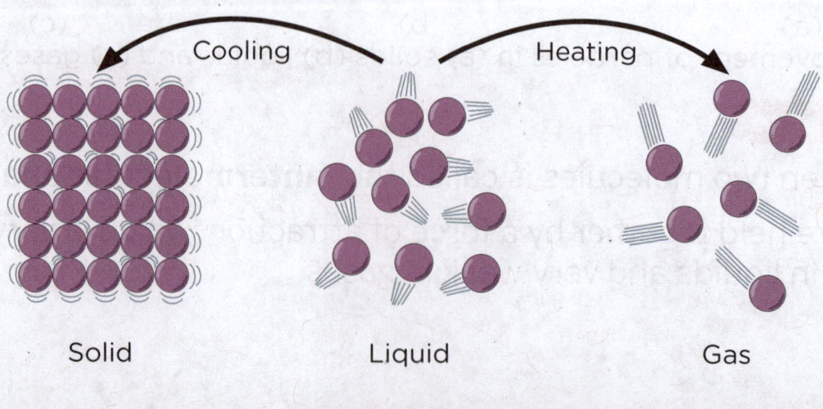

HEATING AND COOLING OF SOLIDS

When solids are heated, the molecules start to vibrate faster. On further heating, the particles move around and gradually turn into a liquid.

This process in which a solid changes into liquid is called **melting**. Not all substances undergo melting at normal conditions. Some solid substances may undergo certain other changes on heating. For example, paper burns when it is heated. However, most solids expand on heating.

Let us understand this concept from the example given below.

Safeena has a metal ball and a ring. Before heating the metal ball, the ball could pass through the ring. Then, she heated the metal ball over a flame until it got very hot. She tried to pass the heated metal ball through the ring and found that the ball could not pass through the ring.

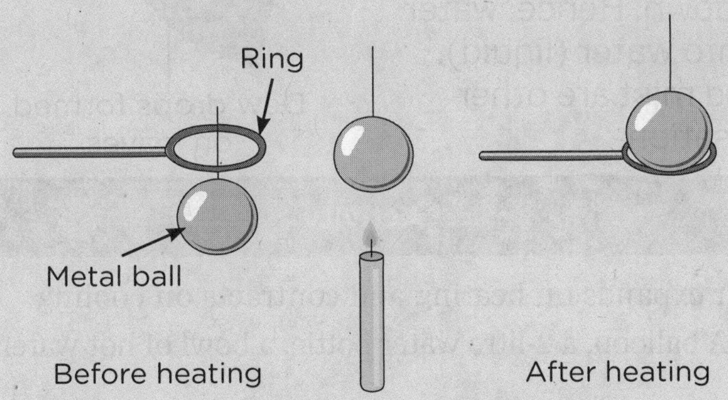

- What would Safeena conclude from this experiment?

  ____

  ____

- How can Safeena pass the metal ball through the ring once again?

  ____

  ____

In the above example, when the ball is heated, the molecules move further apart, causing the size of the ball to increase. This is called **expansion**[1].

When the ball is cooled, the movement of molecules slows down and they come closer to each other. This is called **contraction**[2].

Word Web

[1]**Expansion**: increase in the size of matter

[2]**Contraction**: decrease in the size of matter

HEATING AND COOLING OF GASES

When a gas is heated, the molecules move away very fast, thereby causing it to expand. When a gas is cooled for a long time, the molecules slow down and gradually turn into a liquid. This process in which a gas changes into liquid is called **condensation**.

Water droplets condensing on the surface of the glass

The formation of water droplets on the outer surface of a cold glass is an example of condensation.

When the water vapour present in the air comes in contact with the cold surface of the glass, the movement of molecules in water vapour slows down. Hence, water vapour (gas) turns into water (liquid). Formation of dew and mist are other examples of condensation.

Science in Context

A gas expands faster than a solid or a liquid because the force of attraction between particles of gas is the weakest.

Dew drops formed on leaves

Water vapour condensing to form mist

Activity

Aim: To show that air expands on heating and contracts on cooling

Materials required: A balloon, a 2-litre water bottle, a bowl of hot water and a bowl of cold water

Procedure:

1. Fix a slightly inflated balloon to the mouth of the 2-litre bottle.
2. Place the bottle with the balloon into the bowl of hot water. Observe the change.
3. Place the bottle with the balloon into bowl of cold water and observe the change.

Bottle with balloon in hot water

Bottle with balloon in cold water

Observation: The balloon gets inflated when the bottle is kept in hot water. The balloon gets deflated when the bottle is kept in cold water.

Reason: The hot water heats up the air inside the bottle. The air expands and inflates the balloon. In the bowl of cold water, the air inside the bottle gets cooled and contracts. This causes the balloon to deflate.

Conclusion: Air expands on heating and contracts on cooling.

PHYSICAL AND CHEMICAL CHANGES

A change in which no new substance is formed and there is merely a change of state is called a **physical change**. Melting of ice, freezing of water and boiling of water are examples of physical changes.

Physical changes are reversible. In other words, we can get the original substance back even after the change takes place. For example, ice changes to water on heating, but we can get back ice by freezing water.

A **chemical change** is one in which a new substance is formed. This kind of change is irreversible, that is, we cannot get the original substance back once it undergoes a chemical change. Burning of a wax candle and burning of paper are examples of chemical changes.

Keywords

Atoms: Building blocks of all matter
Molecules: Particles formed when two or more atoms combine
Evaporation: The process in which a liquid changes into gas
Condensation: The process in which a gas changes into liquid
Physical change: A change in which no new substance is formed
Chemical change: A change in which a new substance is formed

Summary

1. A substance that occupies space and has weight is called matter.
2. Matter is made up of atoms. Atoms are very small in size.
3. When two or more atoms combine together they form a molecule.
4. In solids, atoms and molecules vibrate without changing their position. That is why solids have a fixed shape.
5. In liquids, atoms or molecules move around and glide over each other. That is why liquids flow easily and can be poured into containers.
6. In gases, atoms or molecules move freely and are not in contact with each other. That is why gases move to take up the entire space available to them.
7. Matter expands on heating and contracts on cooling.
8. In a physical change, no new substance is formed. Physical changes are reversible.
9. In a chemical change, a new substance is formed. Chemical changes are irreversible.

EXERCISE

A Choose the correct option for each of the following.

1. Matter has mass and occupies
   (a) weight (b) space (c) energy (d) none of these
2. Identify the state of matter in which the molecules spread out and occupy the entire space available.
   (a) Solid (b) Liquid (c) Gas (d) All of these
3. Identify the state of matter that has a fixed shape.
   (a) Solid (b) Liquid (c) Gas (d) Both solid and gas
4. When a solid is heated, it
   (a) expands (b) contracts (c) freezes (d) none of these
5. Formation of mist is an example of ____________ .
   (a) freezing (b) melting (c) evaporation (d) condensation
6. Which of the following is a chemical change?
   (a) Freezing of water (b) Melting of ice
   (c) Burning of paper (d) Boiling of water

B Match the following.

Column A	Column B
(a) Process in which a solid turns into liquid	(1) Condensation
(b) Process in which a liquid changes into solid	(2) Evaporation
(c) Process in which a liquid changes into gas	(3) Freezing
(d) Process in which a gas changes into liquid	(4) Melting

C State whether the following statements are true (T) or false (F).

1. Atoms are the building blocks of matter. ☐
2. Molecules are made up of only the same kind of atoms. ☐
3. When heated, atoms move faster. ☐
4. All substances undergo melting in normal conditions. ☐
5. Solids expand on heating. ☐

D Answer the following in one word.

1. A substance that occupies space and has mass
2. Building block of matter
3. The state of matter in which particles move freely
4. The space between two molecules
5. An example of a liquid

E Answer the following questions briefly.

1. Why is a molecule of oxygen written as O_2?
2. What is the difference between melting and freezing?
3. What happens when a solid is heated?
4. What happens to the particles when a gas is cooled?
5. What is expansion?

F Answer the following questions.

1. Do solids expand on heating? Justify your statement with the help of an activity.
2. Explain what happens when
   (a) a liquid is heated
   (b) a liquid is cooled
3. Explain physical changes and chemical changes with the help of examples.

G HOTS (Higher Order Thinking Skills)

When someone opens a bottle of perfume in one corner of a room, we can smell it from the other corner. Why? (Hint: Think about the movement of particles in gases.)

H Sketch to show how water would look like in each of the containers shown below.

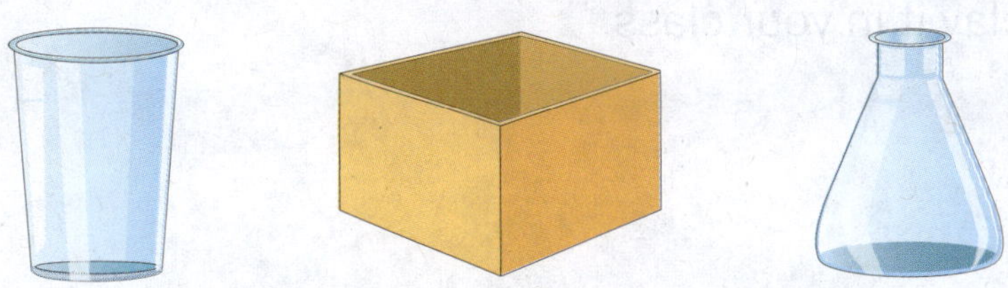

Did the shape of water stay the same or changed in each of the containers?

WORKSHEET

Use the words given in the box to complete the passage given below.

contact	atoms	evaporation	position	freely
motion	attraction	melting	matter	molecule
intermolecular	freezing	building blocks		

Everything in the universe is made up of ____________. Matter is made up of small particles called ____________. Atoms are the ____________ of all matter. When two or more atoms combine together, they form a particle called a ____________. Atoms and molecules of all matter are always in constant ____________. In solids, atoms/molecules move but remain in the same ____________. In liquids, atoms/molecules move around but remain in ____________ with each other. In gases, atoms/molecules move ____________ and are not in contact with each other.

The space between two molecules is called their ____________ space. The molecules are held together by a force of ____________ which is very strong in solids, weak in liquids and very weak in gases.

The process in which a liquid changes into a gas is called ____________.
The process in which a liquid changes into a solid is called ____________.
The process in which a solid changes into liquid is called ____________.

Project

1. Collect pictures of different items found in your household and paste them in your scrapbook. Write solid, liquid or gas below each.
2. Draw the arrangement of particles in solids, liquids and gases on a chart paper and display it in your class.

12 Light and Shadow

Learning Checklist

- Transparent, translucent and opaque materials
- Shadows

Let's Get Started

We have learned in earlier classes that objects that produce light are called **luminous objects**. They help us to see the non-luminous objects, which do not produce light.

In the picture given below, put a tick mark (✓) against luminous objects and a cross mark (✗) against non-luminous objects in the boxes provided.

Non-luminous objects are made from different materials. Materials can be grouped on the basis of the amount of light they allow to pass through them. Let us understand this with the help of an activity.

Activity 1

Aim: To identify different types of materials

Materials required: A clear (see-through) plastic cover, a sheet of butter paper and an A4 size chart paper

Procedure:

1. Take a clear plastic cover, a butter paper and an A4 size chart paper.
2. Ask your friend to stand in front of you.
3. Now, try to see your friend through each—the plastic cover, the butter paper and the chart paper.

Conclusion: You will be able to see your friend clearly through the clear plastic cover, partially through the butter paper and not at all through the chart paper.

TRANSPARENT, TRANSLUCENT AND OPAQUE MATERIALS

Materials that allow light to pass through them completely are called **transparent materials**. Examples of transparent materials are clear glass, clear water and clear plastic sheet.

Materials that allow light to pass through them partially (not completely) are called **translucent materials**. Examples of translucent materials are butter paper, frosted glass and coloured glass.

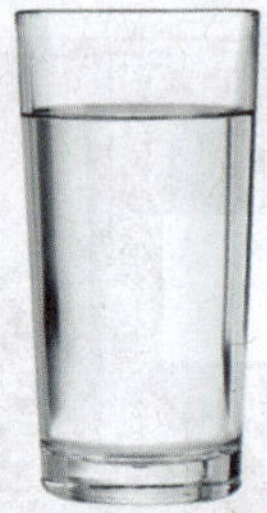
Clear water (transparent)

Frosted glass (translucent)

Know the Fact

Is mirror a transparent object? It is not because we cannot see objects on the other side of a mirror, though we can see our own image clearly.

Materials that block light or do not allow light to pass through them are called **opaque materials**. Examples of opaque materials are wood, bag and books.

Bag (opaque)

Book (opaque)

SHADOWS

We know that a shadow is formed when an object is placed in front of a source of light. Let us perform an activity to understand the formation of shadows by transparent, translucent and opaque objects.

Activity 2

Aim: To understand the formation of shadows by transparent, translucent and opaque materials

Materials required: A transparent plastic sheet (A4 size), a translucent tracing sheet (A4 size), a cardboard sheet (A4 size), a torch and a screen

Procedure:

1. Take the plastic sheet, the tracing sheet, the cardboard sheet and the torch.
2. Hold the objects one by one in front of the screen and project the light from the torch onto each object from the opposite side. What do you observe?

Observation:

1. You will see that the opaque cardboard sheet blocks the light completely and hence forms a clear and dark shadow on the screen.
2. The translucent tracing sheet partially blocks the light and thus forms a lighter shadow on the screen.
3. The transparent glass does not block the light from passing through it and hence does not form any shadow.

From the above activity, it is clear that opaque and translucent objects form shadows, whereas transparent objects do not form shadows.

Thus, a **shadow** is a dark region formed by a translucent or an opaque object when it blocks the path of light. Three things are required for a shadow to form: a light source, a translucent or an opaque object and a screen (a surface on which the shadow can form).

Characteristics of a Shadow

Following are some characteristics of a shadow.

(a) A shadow is always formed on the opposite side of the light source.

(b) A shadow shows only the outline of the shape of the object, and not any other details.

(c) The colour of a shadow does not depend on the colour of the object. Shadows are always dark.

(d) The size of a shadow depends on the distance of the light source from the object and also on the distance of the screen from the object.

When the light source is close to the object or the screen is away from the object, the size of the shadow increases. When the light source is away from the object or the screen is close to the object, the size of the shadow decreases.

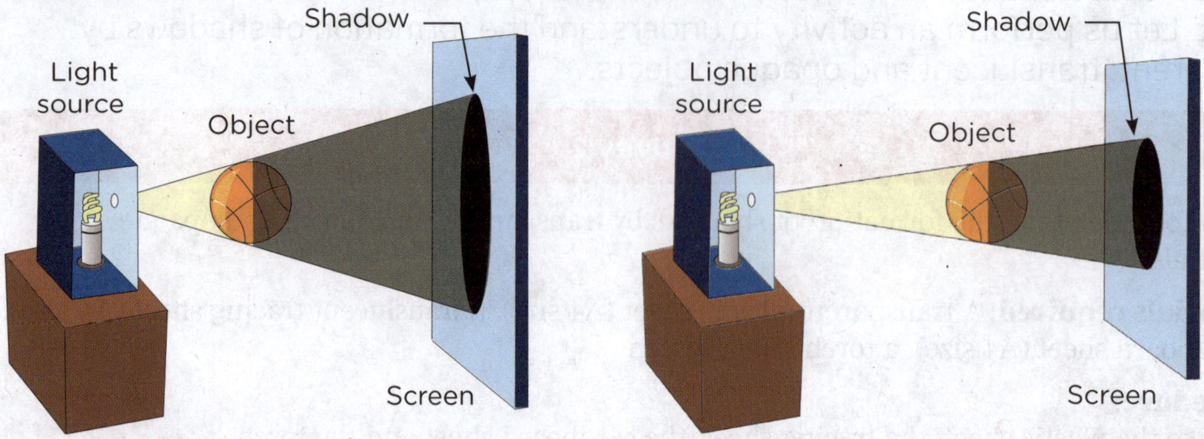

Science in Context

The shadow of an object is the longest in the morning and evening, and the shortest at noon. This is because the rays of the sun are slanting in the morning and evening, and straight above the head at noon.

Let's Say Together

Transparent Translucent Opaque Shadow Screen

Keywords

Transparent materials: Materials that allow light to pass through them completely
Translucent materials: Materials that allow light to pass through them partially
Opaque materials: Materials that do not allow light to pass through them
Shadow: The dark region formed by a translucent or an opaque object when it blocks the path of light

Summary

1. Materials can be classified as transparent, translucent and opaque based on the amount of light they allow to pass through them.
2. Only opaque and translucent objects form shadows.
3. Shadows are always dark.
4. Shadows are formed on the opposite side of the light source.

EXERCISE

A Choose the correct option for each of the following.

1. Objects that give out light are called
   (a) non-luminous (b) opaque (c) luminous (d) transparent
2. Which of the following is a transparent object?
   (a) Clear glass (b) Wood (c) Butter paper (d) Mirror
3. Identify the translucent object from the following.
   (a) Frosted glass (b) Cardboard (c) Mirror (d) Clear water
4. Materials that do not allow light to pass through them are called
   (a) Opaque (b) Transparent (c) Translucent (d) All of these
5. Which of the following objects cannot form a shadow?
   (a) Pencil (b) Clear glass (c) Bag (d) Notebook

B State whether the following statements are true (T) or false (F).

1. Translucent materials allow light to pass through them completely. ☐
2. Shadows are always dark. ☐
3. Only transparent materials can form shadows. ☐
4. Shadows can be formed without a light source. ☐
5. The size of the shadow of an object is always the same. ☐

C Fill in the blanks using the correct words.

1. A shadow is always formed on the ________ side of the object. (same/opposite)
2. Butter paper is an example of ________ material. (transparent/translucent)
3. Shadows are formed by ________ materials. (transparent/opaque)
4. The size of a shadow depends on the distance of the ________ from the light source. (object/shadow)
5. The colour of a shadow is always ________. (white/dark)

D Write two examples for each of the following.

1. Transparent materials ________ ________
2. Translucent materials ________ ________
3. Opaque materials ________ ________
4. Things required to form a shadow ________ ________

E Answer the following question briefly.

1. What is the difference between transparent and translucent materials?
2. Why do transparent materials not form shadows?
3. Which material would form a darker shadow: translucent or opaque? Why?
4. State two factors on which the size of a shadow depends.

F Answer the following questions.

1. Define transparent, translucent and opaque materials. Give two examples of each.
2. What is a shadow? What are the characteristics of a shadow?
3. Why is it that only translucent and opaque materials form shadows?

G HOTS (Higher Order Thinking Skills)

You and your mother went to the market to buy new containers for her kitchen. What kind of containers would you suggest her to buy: opaque, transparent or translucent? Give reasons for your answer. (Hint: Think of the characteristics of these materials.)

H Write transparent, translucent or opaque for each of the following.

Clear water

______

Bag

______

Frosted glass

______

Apple

______

I Project

Change a transparent object into translucent and opaque objects. Also, change an opaque object into a translucent object.

Materials to be used: A4 size transparent sheet of plastic, A4 size tracing paper (2 sheets), A4 size white opaque sheet (2 sheets), small bottle of hair oil, glue

Procedure:

1. Take a transparent plastic sheet. Stick tracing papers on both sides. (Apply glue only on the edges.) Now, the transparent plastic sheet becomes translucent. Paste one of the opaque white sheets on it. Can you see through it now?
2. Take an A4 size opaque white sheet. Place two drops of coconut oil on the sheet and gently spread it across the sheet. You will find that the opaque sheet has become translucent.

Draw the shadows of the tree given below for three different positions of the sun during the day.

Life Skills

You are given the task of setting up your study room. Make a list of the materials you would need. Classify those materials as transparent, translucent and opaque. Justify their purpose and usage. Include two materials from each category.

13 Force and Energy

Learning Checklist

- Effects of force
- Types of force
- Energy
- Simple machines

Let's Get Started

Force and energy are the terms that we often use in our daily lives. To find out how much we know about force and energy, let us do this short exercise.

Look at the picture given below. Fill in the blanks using the correct options.

In a volleyball game, the ball hit by the players initially goes up, but always comes down due to ____________ (frictional/gravitational) force.

The referee on the side uses ____________ (sound/heat) energy from his whistle to control the game.

Since the game is mostly played indoors, the stadium needs to be sufficiently lit. The floodlights produce a lot of ____________ (light/sound) energy.

You have learned in Class 3 that **push** or **pull** are the actions that make objects move, for example, a bullock pulling a cart, a man pushing a bike, a boy lifting his school bag and a girl kicking a ball. In all these cases, what makes the objects move? It is the push or pull applied on the objects.

Force is defined as a push or a pull applied on an object. Let us learn about the effects that force can produce.

EFFECTS OF FORCE

Force can make an object move

A force acting on a stationary object can make it move. Also, a force acting on a moving object can make it move faster, if the force is applied in the same direction in which the object is moving. For example, a football player kicks a ball to make it move. A boy pushes a cart to make it move forward.

Force can move a cart

Force can stop a moving object

A force acting on a moving object can slow it down or stop it completely, if it is applied in the direction opposite to the direction of motion.

For example, a goalkeeper uses his hands to stop a football from rolling into the net.

Force can stop a moving football

Force can change the direction of a moving object

A force acting on a moving object from an angle can change the direction in which the object is moving.

For example, in cricket, a batsman hits a moving ball with such force that it changes its direction.

Force can change the direction of a moving ball

Force can change the shape of an object

A force acting on an object can change its shape.

Have you ever kneaded dough or made chapatis? Notice how a rolling pin makes a ball of dough flat and round when you apply force.

Another example is to squeeze a sponge and see how its shape changes.

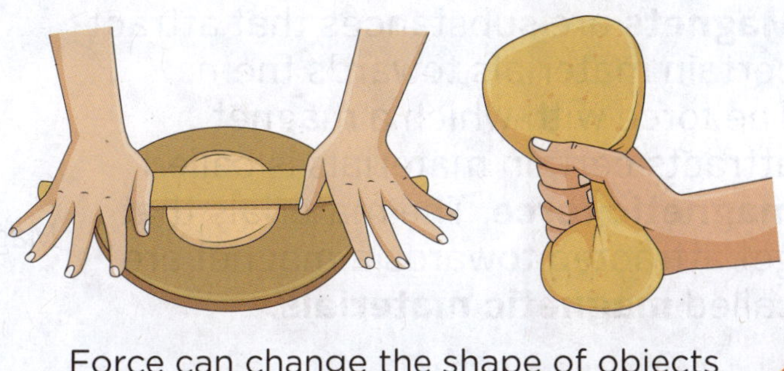

Force can change the shape of objects

TYPES OF FORCE

Various types of forces exist in nature. Some of them are given below.

Gravitational Force

Gravitational force or gravity is the force with which all objects in the universe attract each other. The gravitational force of the Earth pulls objects towards its centre. Any object that is thrown upwards always comes down because of the Earth's gravity. Gravitational force always acts in the opposite direction when an object moves away from the Earth; it acts along the direction of motion of the object when it moves towards the Earth.

Electrostatic Force

The attractive or repulsive force that acts due to the presence of 'charges' is called **electrostatic force**. Let us perform an activity to understand this better.

Activity 1

Aim: To understand electrostatic force

Materials required: Plastic ruler and bits of paper

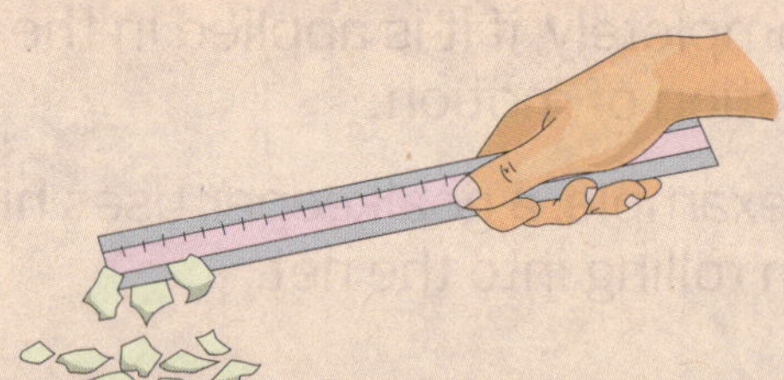

Procedure:

1. Place bits of papers on a table.
2. Take the plastic ruler and rub it against your hair. Ensure that your hair is dry.
3. Bring the plastic ruler near the bits of paper and observe what happens.

Observation: You will find that the bits of paper move towards the plastic ruler.

Conclusion: The plastic ruler gets charged due to rubbing against hair and the bits of paper get attracted towards it when brought closer.

You will learn more about electrostatic force in higher classes.

Magnetic Force

Magnets are substances that attract certain materials towards them. The force with which a magnet attracts certain materials is called **magnetic force**. The materials that get attracted towards a magnet are called **magnetic materials**.

Magnet attracts magnetic materials

Unlike poles attract and like poles repel

Magnetic force

Magnets also apply force on each other. Every magnet has two poles known as the **north pole** and the **south pole**. When two like poles are brought near each other, they repel or oppose each other, whereas unlike poles attract each other.

Frictional Force

Frictional force or **friction** acts on two objects that are in contact and at least one of them is in motion. It always acts opposite to the direction of motion. Hence, it slows down the motion of an object or stops it from moving completely.

The force of friction produced depends on the nature of objects in contact. Rough surfaces offer more friction than smooth surfaces. Let us learn about some advantages and disadvantages of friction.

Advantages of Friction

(a) Friction allows steady motion of objects. For example, we are able to walk steadily due to the friction between the ground and our shoes. The grooves on the sole of our shoes provide enough friction to prevent us from falling. Vehicles move steadily on the road because of the friction between the tyres and the road.

Friction helps us to walk

(b) Friction produces heat, which we use for different purposes. For example, a matchstick is struck against the rough surface of a matchbox, heat is produced due to friction, which lights the matchstick. When we rub our hands together, heat is produced due to friction, which keep our hands warm.

A matchstick is rubbed against a matchbox

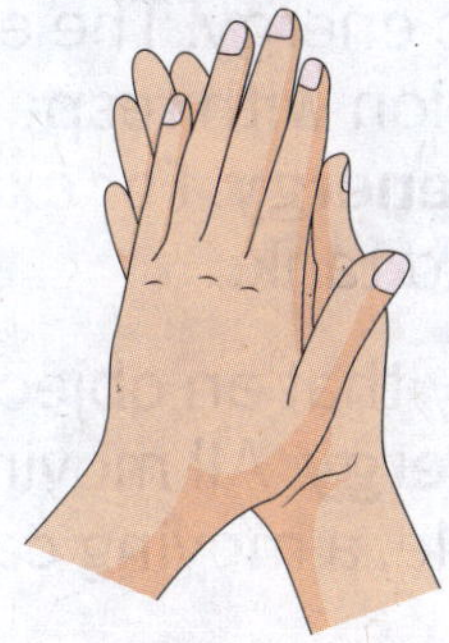
Rubbing hands produces heat

Disadvantages of Friction

(a) With increase in friction, it becomes difficult to move an object as more force is needed to overcome friction. For example, moving heavy objects on a rough surface is difficult.

(b) Friction causes the surface of an object to wear out. For example, soles of shoes and surface of tyres wear out because of friction.

ENERGY

The ability to do work is called **energy**. Energy is present around us in many forms and we get it from different sources. For example, during the day, the sun gives out light and heat energy. At night, street lights use electrical energy to light the roads. The food we eat gives us the energy to work and play.

Let us now learn about the different forms of energy.

Heat Energy

Heat energy

Heat is a form of energy. Heat energy is obtained from the sun, and also by burning fuels like coal, petrol and kerosene. We use heat energy for cooking food and keeping us warm.

Light Energy

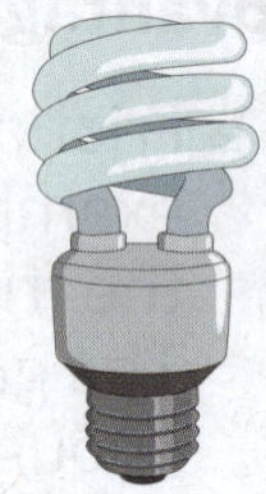

Light energy

Light energy helps us to see things around us. The sun is a natural source of light. Some artificial sources of light are electric bulb, candle, tubelight, lamp, etc. Plants also use light energy from the sun to prepare food.

Mechanical Energy

A moving car has mechanical energy

The energy that an object has due to its position or its motion is called **mechanical energy**.

Mechanical energy is of two types: potential energy and kinetic energy. The energy that an object has due to its position with respect to the ground is called **potential energy**. For example, water stored in a dam or overhead tank.

The energy that an object has due to its motion is called **kinetic energy**. All moving objects have kinetic energy. For example, a moving car and a bullet fired from a gun.

Electrical Energy

Electrical energy

The most common and useful form of energy nowadays is electrical energy. This form can be easily converted to other forms of energy. Sources of electrical energy are cells and generators. Most of the common household appliances such as fans, mobiles, watches, televisions and computers run on electrical energy.

Sound Energy

Sound is a form of energy that is produced by vibration of objects. Some sources of sound energy are radio, loudspeakers and musical instruments.

Speaker

Wind Energy

The energy obtained from the moving wind is called wind energy. Windmills help in converting wind energy into electrical energy.

Windmill

SIMPLE MACHINES

Machines are tools that make our work easier by reducing the amount of force (effort) needed to do the work. Machines can be simple or complex.

Tools such as scissors, screws, tongs and knives that help us in doing work efficiently and easily are called **simple machines**. There are six basic types of simple machines: lever, pulley, inclined plane, screw, wedge, and wheel and axle.

Complex machines are made of two or more simple machines. For example, a bicycle is a complex machine made up of lever, screw, and wheel and axle.

Let us learn more about simple machines.

Lever

A lever is a rigid rod made to rest on and move around a fixed support. It consists of three parts:

Fulcrum: It is the fixed point or support on which the rod moves freely.

Effort: It is the force that is applied on the free end of the rod to enable the other end to perform work.

Load: It is the object on which work has to be done.

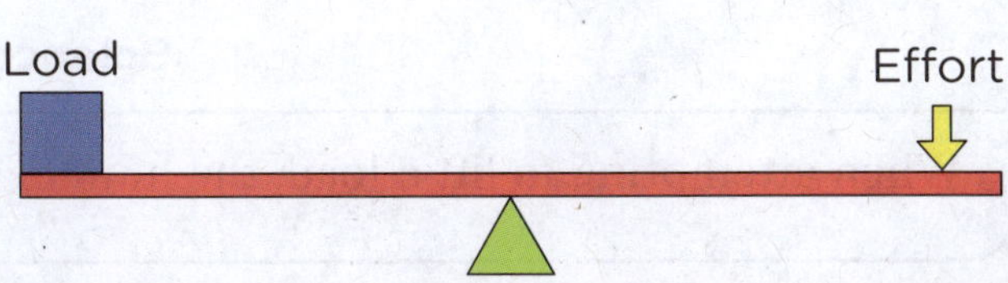

Depending on the position of these parts, levers are classified into three types: first-class, second-class and third-class.

First-class levers: In this type of levers, the fulcrum is between the load and the effort. Some examples are scissors, crowbars, see-saw and claw hammers.

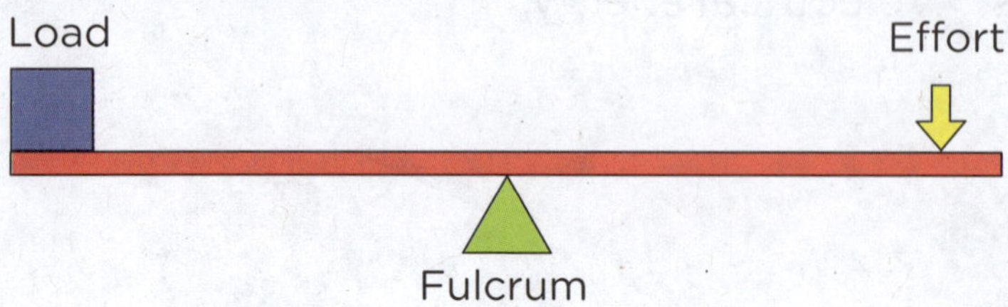

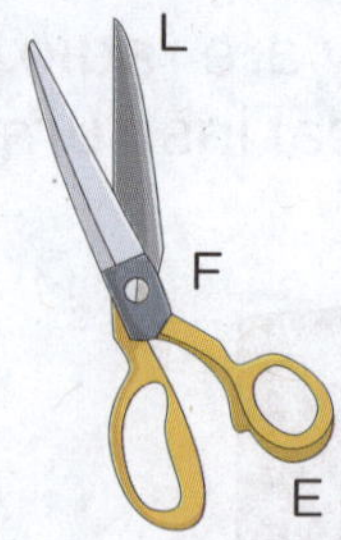

Scissors

See-saw

Second-class levers: In this type of levers, the load is between the effort and the fulcrum. Some examples are wheelbarrows, bottle openers and lemon squeezers.

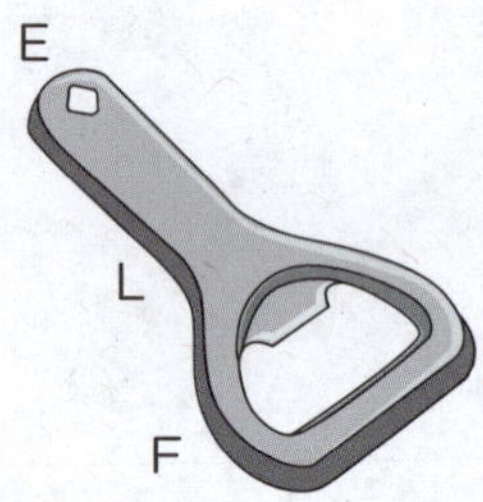

Bottle opener

Wheelbarrow

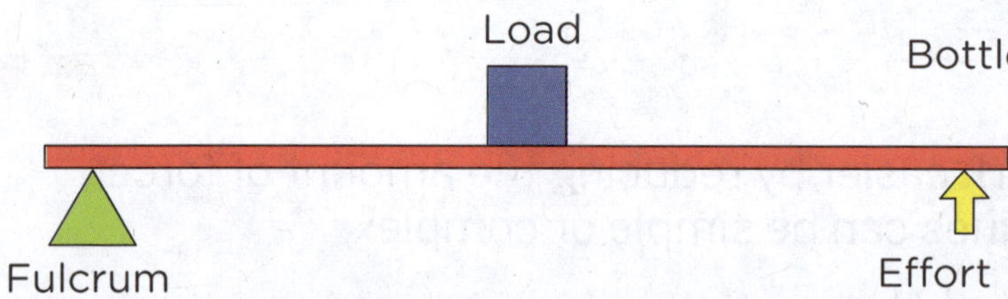

Third-class levers: In this type of levers, the effort is between the fulcrum and the load. Some examples are a pair of tongs, a stapler, a fishing rod and tweezers.

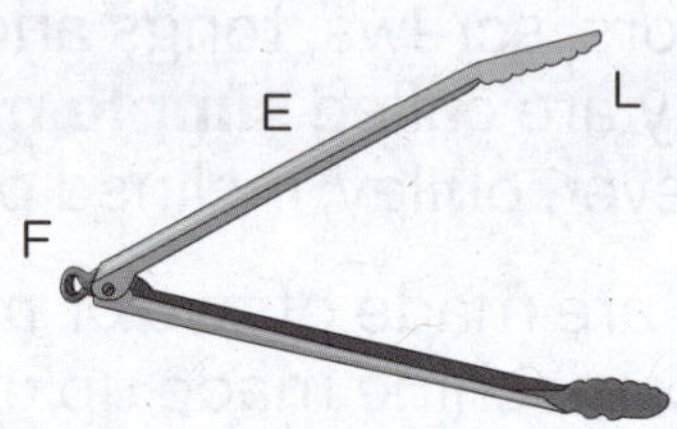

Tongs

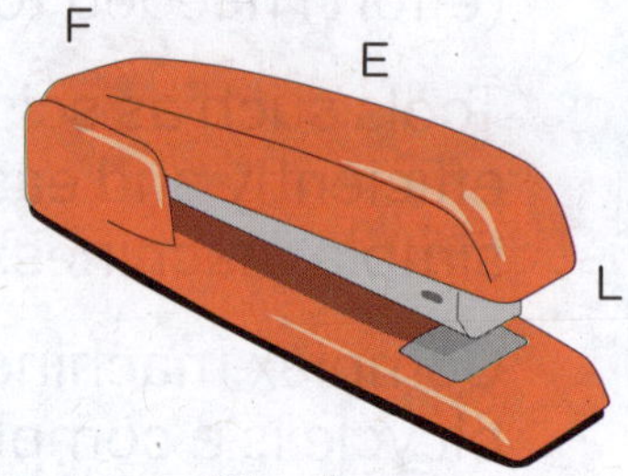

Stapler

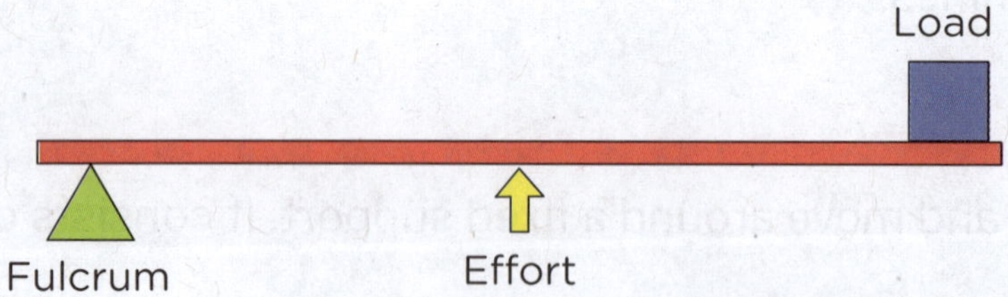

Science in Context

When attempting to lift a load, the human hand acts as a third-class lever.

Wheel and Axle

Wheel and axle is a simple machine in which a wheel is attached to a rod called the axle. The wheel and the axle work together. When force is applied to the wheel, the axle turns. On the other hand, if we apply force to the axle, the wheel turns. Some examples are steering wheels, door knobs and giant wheels.

Giant wheel

Activity 2

Aim: To study the wheel and axle combination

Materials required: Four bottle caps, two pieces of thick broomstick (6 cm length), two straws (4 cm length), the matchstick holder of a matchbox and glue

Procedure:

1. Take the broomsticks and insert them into the straws.
2. Fix the bottle caps to the ends of the broomsticks to look like a wheel and axle.
3. Insert the broomsticks into the matchstick holder as shown in the image given alongside.
4. Place a weight on top of the matchstick holder and push the set-up from behind.

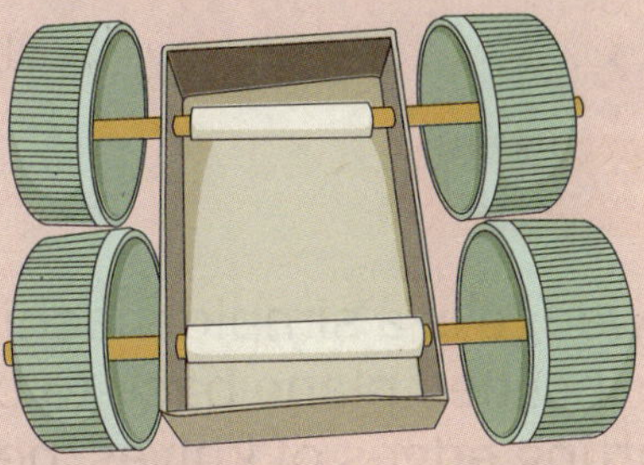

Observation: The set-up moves forward with the help of wheel and axle.

Conclusion: Wheel and axle combination makes work easier by reducing effort.

Pulley

A pulley is made up of a small wheel with a groove on its outer edge and a rope or chain running over the groove. The wheel rotates around an axle. They are used to lift weights. They are of two types: fixed and movable.

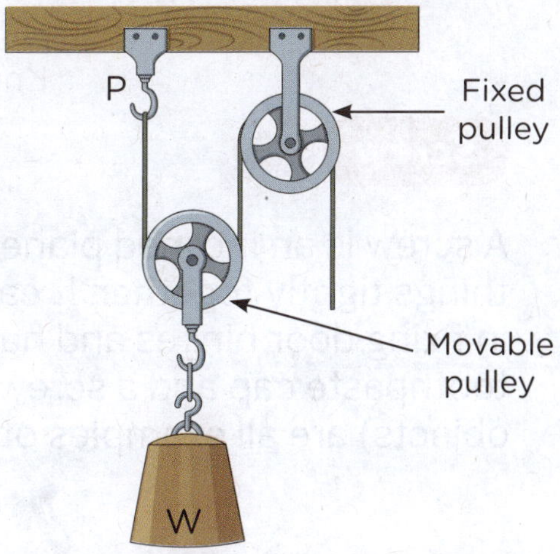

A pulley

In a fixed pulley, the axle is attached to a support like a hook or wall. It makes work easier by changing the direction of the force applied. Some examples are pulleys used in wells and on flagpoles.

In a movable pulley, the axle is not attached to any support and is free to move along with the load. Movable pulleys are used along with fixed pulleys to reduce the effort. An example is cranes used in construction sites to lift heavy loads with less effort.

Inclined Plane

An inclined plane is a slope or a ramp or any slanting surface that helps to carry loads. It requires less force to move an object up an inclined plane than it does to lift the object vertically. Some examples are a slide in the park, ramps in hospitals to push wheelchairs up and down, and ramps for loading and unloading goods on trucks and aeroplanes.

A slide

Ramp for loading goods

Wedge

A wedge is a simple machine made by placing two surfaces at least one of which is an inclined plane back to back to form a sharp cutting surface. Some examples are cutting edges of knives, needles and axes.

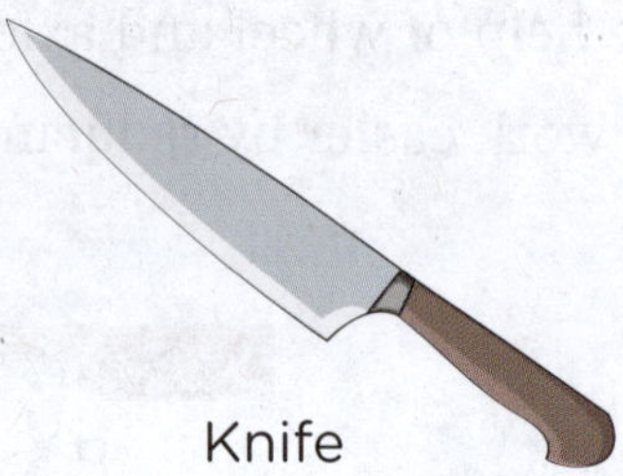
Knife

Axe

Screw

A screw is an inclined plane wrapped around a central cylinder. It is used to hold things tightly together. It can be turned using a screwdriver. Screws are useful in fixing door hinges and handles. Nuts and bolts, a dentist's chair, a bottle or toothpaste cap and a screw jack (big screws used to lift cars and other heavy objects) are all examples of screws.

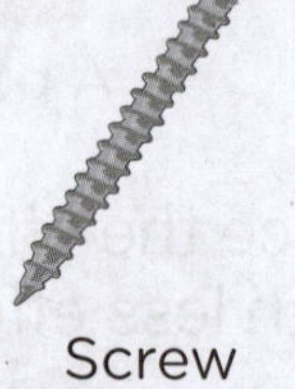
Screw

Bottle cap

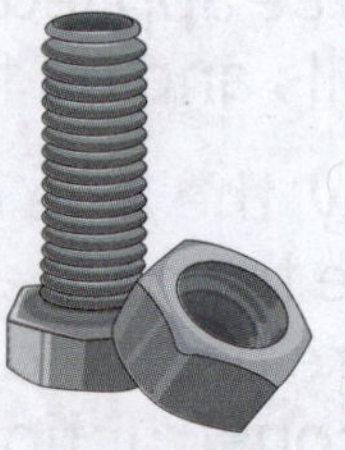
Nut and bolt

Activity 3

Aim: To make a screw from an inclined plane

Materials required: A pencil, a sketch pen and a triangular sheet of paper

Procedure: Colour the inclined surface of the triangular paper. Now, wrap this paper around the pencil so that the coloured surface faces outside.

Observation: The coloured surface resembles a winding road or a staircase. We have thus made screw out of an inclined plane.

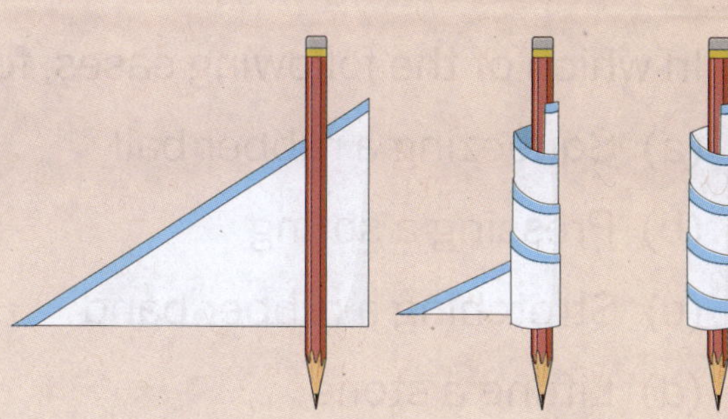
a

Let's Say Together

Electrostatic Friction Potential energy Kinetic energy

Keywords

Magnetic force: The force with which a magnet attracts certain materials

Frictional force: A force that acts on two objects that are in contact and at least one of them is in motion

Heat energy: Energy obtained by burning fuels like oil and gas

Mechanical energy: Energy that an object has due to its position or motion

Sound energy: Energy produced by the vibration of objects

Simple machine: A tool or a device that makes work easier by reducing the effort

Summary

1. Force is a push or a pull applied on an object.
2. Force can start or stop the motion of an object or change the speed and direction of motion or set an object in motion at rest.
3. Force can change the shape of an object.
4. The different types of forces are gravitational, frictional, magnetic and electrostatic.
5. Heat, light, electrical, mechanical, wind and sound are different forms of energy.
6. Machines make our work easier.
7. Six types of simple machines are lever, wheel and axle, inclined plane, pulley, wedge and screw.

EXERCISE

A Choose the correct option for each of the following.

1. In which of the following cases, force does not change the shape of an object?
   (a) Squeezing a rubber ball
   (b) Pressing a spring
   (c) Stretching a rubber band
   (d) Lifting a stone
2. Which of the following is a type of force?
   (a) Electrostatic (b) Light (c) Sound (d) Heat
3. Which of the following is an example of a third-class lever?
   (a) Wheelbarrow (b) Tongs (c) See-saw (d) Bottle opener
4. Which of these forces is always attractive in nature?
   (a) Friction (b) Electrostatic (c) Gravity (d) Magnetic
5. Which of the following is not a simple machine?
   (a) Lever (b) Pulley (c) Wheel and axle (d) Bicycle

B State whether the following statements are true (T) or false (F).

1. Force can change the direction of motion of an object. ☐
2. Energy and force are one and the same. ☐
3. Sun is a source of only heat energy. ☐
4. Simple machines reduce the effort required to do work. ☐
5. Pulley does not change the direction of the force applied. ☐

C Choose the odd one out.

1. Scissors, crowbar, see-saw, bottle opener
2. Stapler, wheelbarrow, tweezers, tongs
3. Ramp, flyover, slide, see-saw
4. Screw, bottle cap, ramp, nut and bolt
5. Knife, needle, flagpole, axe

D Fill in the blanks using the correct option.

1. Friction always acts ____________ the direction of motion. (opposite to/along)
2. A fan uses ____________ energy to work. (electrical/light)
3. A light bulb gives ____________ energy. (light/sound)
4. A wheelbarrow is an example of a ____________ lever. (second-class/third-class)
5. A simple machine used to draw water from a well is ____________. (lever/pulley)

E Give two examples for each of the following.

1. Sources of sound energy ____________ ____________
2. Third-class lever ____________ ____________
3. Pulley ____________ ____________
4. Screw ____________ ____________
5. Wedge ____________ ____________

F Answer the following questions briefly.

1. Mention the effects of force on objects.
2. What is friction?
3. What is energy?
4. What is a simple machine?
5. Name the six types of simple machines.
6. What is fulcrum?
7. What is a wedge?

G Answer the following questions.

1. Define friction. What are its advantages and disadvantages?
2. What is a lever? Explain the types of levers with one example of each.

H HOTS (Higher Order Thinking Skills)

1. When you stand out in the sun, which energies would you be affected by?
2. How are nails different from screws? Which one will you use to hold things tightly?
3. How are a screw, a wedge and an inclined plane interrelated?

I **Identify the energy represented in each of the images given below.**

1.

____

2.

____

J **Project**

1. Look at a bicycle and try to identify the simple machines that are used to make it function. Write them in the table given below.

Name of the cycle part	Simple machine represented

2. Identify the simple machines used in your house and make a list of them.

Identify the six simple machines shown in the picture given below. Write the name and the type of each in the space provided.

1. ______________________________

2. ______________________________

3. ______________________________

4. ______________________________

5. ______________________________

6. ______________________________

WORKSHEET

14 Measurement

Learning Checklist

- History of measurement
- Measurement of length, capacity, mass, time and temperature

Let's Get Started

Finding out the size or quantity of something is called **measurement**. We measure many things in our day-to-day lives. Look at the pictures given below and complete the sentences.

(1)

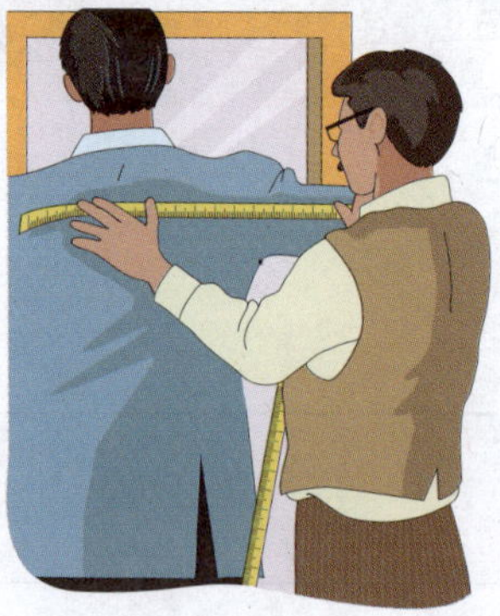

(2)

1. The tailor is measuring the ______________ of the man's shoulder. (weight/length)
2. The greengrocer is ______________ the vegetables. (weighing/carrying)

(3)

(4)

3. The milkman is measuring the ______________ of milk. (height/quantity)
4. The doctor is measuring the ______________ of the patient. (weight/temperature)

Have you ever wondered how people in ancient times used to measure things? What do you think were the methods used by them? Let us look into the past and find out.

HISTORY OF MEASUREMENT

In the earlier days, people used their body parts for measuring length.

Some of the body parts that have been used to measure length are hand span, pace, cubit, foot and thumb.

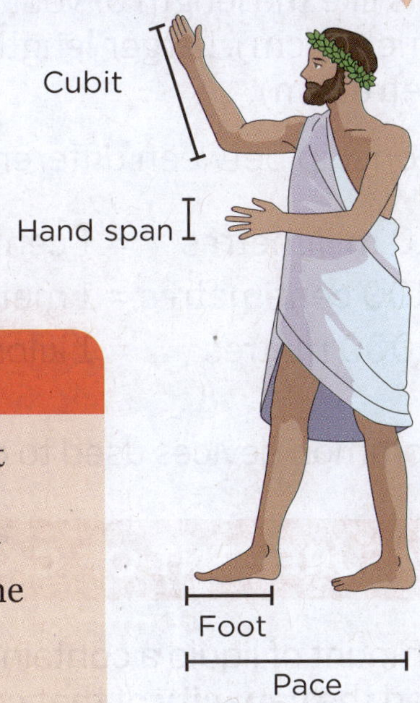

Body parts used for measurement in earlier days

Activity 1

Aim: To check the reliability of body parts for measurement

Procedure: Compare your hand span, cubit and foot size with that of your classmate and teacher, and find out if they are the same, smaller or bigger. Write your observation in the table given below.

Observation:

Person	Hand span	Cubit	Foot
You			
Your friend			
Your teacher			

Conclusion: The size of the body parts vary from person to person and hence the measurements taken using them will also vary.

From Activity 1, it is clear that the size of body parts differs from person to person. Hence, measurement using body parts does not give accurate results. This led scientists to develop standard units for measurement, which would remain constant across the globe.

A fixed standard used for measurement is called **unit**. Length, mass, capacity, time and temperature are the different quantities we measure. We now have fixed units for the measurement of these quantities.

MEASUREMENT OF LENGTH

Have you ever been to a shop to buy unstiched cloth material? The shopkeeper measures the cloth material with a long steel rod. It is a measuring rod that measures **length** in metres. The standard unit of measurement of length is metre (m). Smaller lengths like the length of your notebook are measured in millimetre (mm) and centimetre (cm). Larger lengths like the distance between two cities are measured in kilometre (km).

Relationship between different units for measuring length:

10 millimetres	= 1 centimetre
100 centimetres	= 1 metre
1000 metres	= 1 kilometre

A ruler is used to measure small lengths

The common devices used to measure small lengths are a measuring tape and a ruler.

MEASUREMENT OF CAPACITY

The amount of liquid a container can hold is called the **capacity** of that container. The standard unit for measuring capacity is litre (l). We use millilitre (ml) for measuring smaller capacities and kilolitre (kl) for larger capacities. For example, a milkman gives milk in litres. At the hospital, a dropper is used to measure medicines in millilitres.

Relationship between different units of measurement of capacity:

1000 millilitres	= 1 litre
1000 litres	= 1 kilolitre

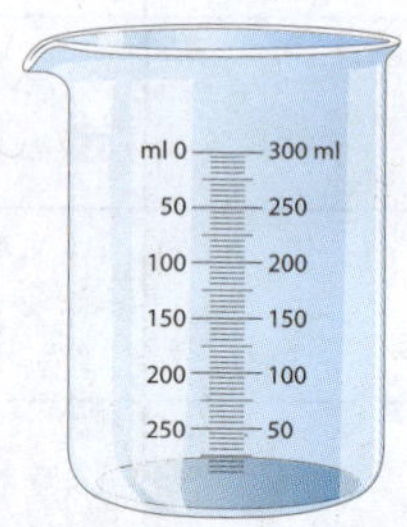

A beaker is used to measure the quantity of a liquid

The capacity of a water tank is much more than that of a beaker

Objects like beakers and measuring cylinders are used to measure capacity.

MEASUREMENT OF MASS

The **mass** of a body or an object is the amount of matter present in it. The standard unit of measurement of mass is kilogram (kg). The smaller units of mass are milligram (mg) and gram (g). For example, you buy potatoes and tomatoes from the greengrocer in kilograms.

Relationship between different units of measurement of mass:

A beam balance

1000 milligrams	= 1 gram
1000 grams	= 1 kilogram

The commonly used device for measuring mass is a physical balance (or beam balance).

MEASUREMENT OF TIME

A particular moment in a day is called **time**. Time period is the duration between two events.

Measurement of time helps us to manage the activities in our daily lives effectively. For example, we know at what time we should wake up in the morning to be in time for school. When we know the timings of events and activities that are going to take place, we can manage our daily work accordingly.

A day is divided into 24 equal durations and each duration is called an hour (h). Smaller durations are measured in minutes (min) and seconds (s), and larger durations are measured in weeks, months and years. The standard unit of measurement of time is second.

Relationship between different units of measurement of time:

60 seconds	= 1 minute
60 minutes	= 1 hour
24 hours	= 1 day
7 days	= 1 week
12 months	= 1 year

A wristwatch

A wall clock

A stopwatch

The commonly used devices to measure time are watches, wall clocks and stopwatches.

Science in Context

A year contains 365.25 days. It is difficult to represent 0.25 days every year. Therefore, it has been agreed globally to add one day to the month of February every fourth year, which is called a **leap year**. A leap year has 366 days, whereas a normal year has 365 days.

MEASUREMENT OF TEMPERATURE

The degree of hotness or coldness of any object is called its **temperature**.

Temperature is commonly measured using the units degree Celsius (°C) and degree Fahrenheit (°F). The standard unit of measurement of temperature is kelvin (K).

Relationship between different units of measurement of temperature:

$$°C = \frac{5}{9}(°F - 32)$$

$$°F = \frac{9\,°C}{5} + 32$$

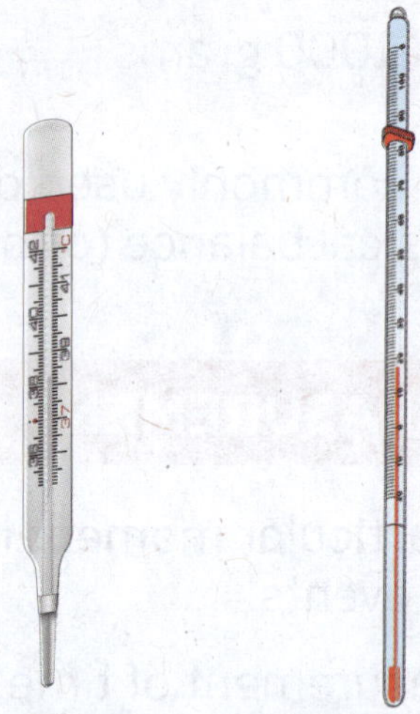

A clinical thermometer   A laboratory thermometer

The commonly used device for measuring temperature is called a **thermometer**.

Clinical thermometers are used in hospitals for measuring body temperature. Has a doctor ever put a thermometer in your mouth and asked you to keep still for a minute or so? The doctor was actually checking you for fever. In other words, the doctor was checking how hot your body is. A **laboratory thermometer**, on the other hand, is used in a science laboratory for measuring temperature of substances during experiments.

Science in Context

Water boils at 100 °C.
Ice melts at 0 °C.
Normal temperature of the human body is 98.4 °F (37 °C).

Let's Say Together

unit   metre   litre   kilogram   degree Celsius   kelvin

Keywords

Measurement: Finding out the size or quantity of something
Unit: A fixed standard used for measurment
Mass: The amount of matter present in a body or an object
Capacity: The amount of liquid a container can hold
Time: A particular moment in a day
Time period: The duration between two events
Temperature: The degree of hotness or coldness of any object

Summary

1. In earlier days, people used their body parts for measuring length, mass, capacity, time and temperature.
2. Some of the body parts used during early days for measurement of length are hand span, cubit, foot and thumb.
3. Length, mass, capacity, time and temperature are the different quantities we measure.
4. The standard unit of measurement of length is metre; the other units are millimetre, centimetre and kilometre.
5. The standard unit of measurement of mass is kilogram; the other units are milligram and gram.
6. The standard unit of measurement of capacity of liquids is litre; the other units are millilitre and kilolitre.
7. The standard unit of measurement of time is second; the other units are minute, hour, week, month and year.
8. The standard unit of measurement of temperature is kelvin; the other units are degree Celsius (°C) and degree Fahrenheit (°F).

EXERCISE

A Choose the correct option for each of the following.

1. Identify the unit of length which is not related to a human body part.
   (a) Cubit (b) centimetre (c) Foot (d) Hand span
2. Which among the following is the biggest unit of length?
   (a) metre (b) millimetre (c) kilometre (d) centimetre
3. How many millilitres make one litre?
   (a) 1000 (b) 100 (c) 10 (d) 1
4. Which among the following is the smallest unit of time?
   (a) second (b) hour (c) minute (d) day
5. Degree of hotness and coldness of an object is called
   (a) Capacity (b) Time (c) Mass (d) Temperature

B Fill in the blanks using the correct option.

1. Fixed standards used for measurement are called __________ . (cubit/units)
2. 1 kilometre is equal to __________ metres. (100/1000)
3. The amount of matter present in an object is called its __________ . (mass/capacity)
4. 1 hour is equal to __________ minutes. (24/60)
5. A stopwatch is used to measure __________ . (temperature/time)

C State whether the following statements are true (T) or false (F).

1. One centimetre is longer than one kilometre. ☐
2. Litre is a unit of mass. ☐
3. A day is divided into 24 equal durations. ☐
4. Clinical thermometers are used by doctors. ☐
5. Laboratory thermometers are used in a science laboratory. ☐

D Name the following.

1. The standard unit for measurement of length __________
2. The standard unit for measurement of capacity __________
3. The standard unit for measurement of mass __________
4. The standard unit for measurement of time __________
5. The standard unit for measurement of temperature __________

E Answer the following questions briefly.

1. What is measurement?
2. Name two devices used to measure small lengths.
3. What is capacity?
4. Define time and time period.
5. What is the relationship between degree Celsius and degree Fahrenheit?

F Answer the following questions.

1. Name three body parts used in earlier days for measuring length. What is the drawback of using these body parts for measuring length?
2. What are the five different quantities of measurement? Also, mention the standard units and devices used for measuring them.
3. Define temperature. Name the various units used to measure temperature. What is the device used for measuring temperature?

G HOTS (Higher Order Thinking Skills)

1. Normal school timing is approximately 7 hours a day. Express it in terms of minutes. (Hint: Remember 1 hour is equal to 60 minutes.)
2. Rohan wants to measure the following quantities, but he is confused about the units to use. Can you help him?
   (a) Length of his tiffin box   (b) Capacity of his water bottle
   (c) Weight of his school bag

H Write the quantities measured using the devices given below.

(a)

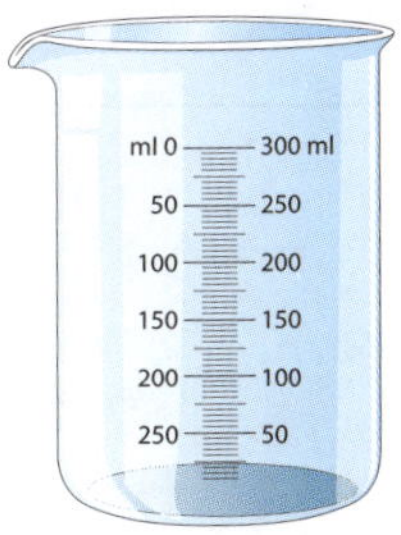

Beaker ____________________

(b)

Beam balance ____________________

(c)

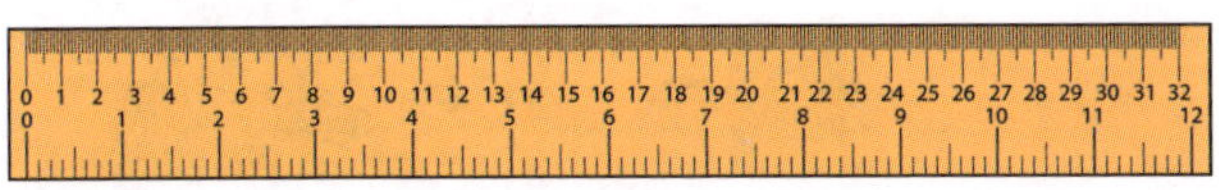

Ruler ____________________

(d)

Clinical thermometer ____________________

I Project

Measure the length of the teacher's table in your classroom using your hand span and then using a scale. Record your readings in the table given below.

Quantity Measured	Measurement Using Hand span	Measurement Using Scale
Length of teacher's table		

Compare your readings with that of your classmates' and find out in which of the cases you got a uniform reading of measurement.

J Life Skills

Rohan has a fever and his mother wanted to check his body temperature. She pulled out a laboratory thermometer from his father's cupboard and tried to measure his temperature. Even after multiple attempts, she was not able to get the correct temperature.

Why was Rohan's mother unable to measure his body temperature correctly?

Choose the most appropriate unit for the measurement of the following quantities given below.

a. litre or millilitre?

Capacity of a glass tumbler

____

Capacity of a water tank

____

b. centimetre or kilometre?

Length of a pencil

____

Distance between Delhi and Kolkata

____

c. hours or years?

Time between sunrise and sunset

____

Chintu's age

____

ASSESSMENT 1

Chapters 1—4

(Maximum Marks = 25)

A Choose the correct option for each of the following. **(7 × 1 = 7)**

1. Embryo, cotyledon and hilum are found in
   (a) Fruit (b) Seed (c) Flower (d) Stem
2. Which of the following is a common feature in dandelion, maple and cotton plants?
   (a) They grow to the same height (b) They have same coloured leaf
   (c) They have round and spiky seeds (d) They have hairy and wing-shaped seeds
3. Which of the following has scales as its body covering?
   (a) Snail (b) Snake (c) Sheep (d) Swan
4. Which of the following animal breathes with the help of gills?
   (a) Earthworm (b) Cockroach (c) Dolphin (d) Prawn
5. Which of the following has a hinge joint?
   (a) Knee (b) Shoulder (c) Wrist (d) Foot
6. The other name of the bony box that protects the brain is
   (a) Cranium (b) Cervix (c) Rib (d) Spine
7. Which of the following helps in the growth and repair of tissues?
   (a) Carbohydrates (b) Proteins (c) Vitamins (d) Fats

B State whether the following statements are true (T) or false (F). **(6 × 1 = 6)**

1. Cotyledon is the source of food for the developing baby plant.
2. Berries are dispersed by wind.
3. Fish swim with the help of their webbed feet.
4. Desert habitats are dry and sandy lands with scanty rainfall.
5. Hinge joint is located between the skull and the backbone.
6. Vegetables and fruits are good sources of roughage.

C Replace the underlined word in each of the following sentences with the correct word. **(6 × 1 = 6)**

1. <u>Reproduction</u> is the process by which a plant grows from a seed.
2. A plant having a hook-like structure on the seeds is <u>sandspur</u>.
3. <u>Lungs</u> are the breathing organs of insects.
4. <u>Pivot</u> joint is present in the elbows, knees, fingers and toes.
5. Anaemia is caused due to the deficiency of <u>vitamin A</u>.
6. A diet containing the right proportions of various nutrients is known as <u>chemical</u> diet.

D Name the following. **(6 × 1 = 6)**

1. The structure that protects the cotyledons and the baby plant inside a seed
2. The largest habitat of the world
3. The tough elastic tissue that is present in the ear and the nose
4. The muscles that are under our control
5. The process of using vaccines for the prevention of diseases
6. The state of complete physical and mental wellbeing

ASSESSMENTS

ASSESSMENT 2

Chapters 5—7

(Maximum Marks = 25)

A Pick the odd one out from each of the following groups of words. **(7 × 1 = 7)**

1. Sprain	Snake bite	Fracture	Sling
2. Fracture	Bee sting	Rabies	Snake bite
3. Troposphere	Biosphere	Mesosphere	Stratosphere
4. Sedimentation	Distillation	Evaporation	Chlorination
5. Ozone	Oxygen	Nitrogen	Meteors
6. Earth	Venus	Mars	Moon
7. Mantle	Core	Corona	Crust

B Circle YES if the statement is correct and NO if it is wrong. **(6 × 1 = 6)**

1. Ganesh fractured his leg. Should he run around and play for the quick healing of the fracture? Yes No
2. Should we tie a tight elastic bandage just above the blisters formed due to burns? Yes No
3. Is it a good practice to give first aid before taking the patient to the hospital? Yes No
4. Distilled water is the purest form of water. Yes No
5. Earth's atmosphere protects us from the harmful radiations of the sun. Yes No
6. Moon has a thick atmosphere around it. Yes No

C Fill in the blanks using the correct words. **(6 × 1 = 6)**

1. A person who has been injured or attacked is called a ________________. (culprit/victim)
2. A strap or cloth that goes under the arm and then around the neck to support the fractured arm is called ________________. (sling/scarf)
3. In water purification, alum is added during the process of ________________. (sedimentation/distillation)
4. ________________ is a process by which plants prepare their own food. (Respiration/Photosynthesis)
5. The innermost layer of the Earth is known as ________________. (core/mantle)
6. The surface temperature of sun is about ________________ degree Celsius. (600/6000)

D Name the following. **(6 × 1 = 6)**

1. The device used to put out fire
2. The amount of water present in air
3. The chemical used to increase the rate of sedimentation
4. The eclipse that occurs when the moon comes between the sun and the Earth
5. The layer from which magma comes out to the surface of the Earth during volcanic eruptions
6. An object moving around a planet in the solar system

ASSESSMENT 3

Chapters 8—11

(Maximum Marks = 25)

A Choose the correct option for each of the following. **(7 × 1 = 7)**

1. Global warming is caused due to the increase in the level of
   (a) Oxygen (b) Water (c) Carbon dioxide (d) Dust
2. Which of the following is a major cause of soil erosion?
   (a) Oil spill (b) Vehicles (c) Global warming (d) Deforestation
3. Hot magma reaching the Earth's surface is called
   (a) Vent (b) Lava (c) Volcano (d) Focus
4. Which of the following is not an after-effect of floods?
   (a) Loss of life (b) Damage to property
   (c) Change in weather pattern (d) Spread of disease
5. Which among the following is an igneous rock?
   (a) Obsidian (b) Sandstone (c) Limestone (d) Slate
6. The rock that contained pebbles and boulders held together by minerals.
   (a) Conglomerate (b) Sandstone (c) Pumice (d) Slate
7. The formation of mist is an example of
   (a) Freezing (b) Melting (c) Evaporation (d) Condensation

B State whether the following statements are true (T) or false (F). **(6 × 1 = 6)**

1. Global warming melts the ice in the polar regions.
2. Asthma is caused due to water pollution.
3. Tides reach the maximum height when both sun and earth are in line with each other.
4. A vent is a region where the Earth's crust is weak and develops a crack.
5. Marble and slate are the examples of metamorphic rocks.
6. All solids contract on cooling.

C Replace the underlined word in each of the following sentences with the correct word. **(6 × 1 = 6)**

1. <u>Noise</u> pollution is a major cause of diarrhoea.
2. Soil erosion can be prevented by <u>deforestation</u>.
3. Intensity of an earthquake is determined by using an instrument called <u>barometer</u>.
4. <u>Marble</u> is formed due to the slow cooling of magma.
5. <u>Oxygen</u> is the major constituent of air.
6. Ozone layer is found in the region of atmosphere called <u>exosphere</u>.

D Answer each of the following questions in one word. **(6 × 1 = 6)**

1. The removal of topsoil by wind and water
2. The types of volcanoes that have not erupted for a long period but have the potential to lead to a possible eruption
3. The rock that is also called volcanic glass

4. The non-metallic mineral commonly used in thermal power generation
5. The gas used during the process of photosynthesis
6. The chemical used to kill germs during water purification

ASSESSMENT 4

Chapters 12—14

(Maximum Marks = 25)

A Pick the odd one out from each of the following groups of words. **(7 × 1 = 7)**

1. Butter paper	Plain glass	Tracing paper	Frosted glass
2. Leather jacket	Silver plate	Butter paper	Leaf
3. Plain water	Plain glass	Frosted glass	Plain plastic
4. Bulb	Candle	Torch	Guitar
5. Hand span	Cubit	Foot	Metre
6. Scissors	Bottle opener	Knife	Tongs
7. Scale	Inch tape	Beam balance	Human foot

B Circle YES if the following statement is correct and NO if it is wrong. **(6 × 1 = 6)**

1. All opaque objects form shadow. Yes No
2. All transparent objects form shadow. Yes No
3. Force can set a stationary object into motion. Yes No
4. Gravitational force is a repulsive force. Yes No
5. One hour is equal to 3600 seconds. Yes No
6. Measuring jar is a measuring device for mass. Yes No

C Fill in the blanks with the correct words. **(6 × 1 = 6)**

1. Shadows are always ________________ in colour. (dark/white)
2. Spectacles are made of ________________ material. (transparent/translucent)
3. The ________________ force is both attractive and repulsive in nature. (magnetic/gravitational)
4. A ________________ surface will offer more friction. (smooth/rough)
5. A day is divided into ________________ hours. (24/30)
6. The SI unit of measurement of temperature is ________________ . (degree Celsius/kelvin)

D Answer each of the following questions in one word. **(6 × 1 = 6)**

1. The materials that allow light to pass through them partially
2. The force that comes into play when the surfaces of two objects are in contact with each other
3. The energy produced by vibrating bodies
4. The device used for measuring temperature
5. A unit for measuring mass other than kilogram
6. One litre expressed in millilitre

COMPREHENSIVE ASSESSMENT 1

Chapters 1—7

(Maximum Marks = 50)

A Name the following. **(10 × 1 = 10)**

1. The reproductive structures in ferns
2. The natural surroundings where living organisms live
3. The structures that attach muscles to the bones
4. The type of the joint present in elbows, knees, fingers and toes
5. The type of disease that is passed from an infected person to a healthy person
6. The immediate treatment given to an injured person before taking him/her to a hospital
7. The process through which the oxygen content of water is increased
8. The layer where most of the meteors burn up while entering the Earth's surface
9. The process in which a pure liquid is separated from a mixture of liquids
10. The outer layer of the sun

B In each of the following questions, two choices are correct. Identify the correct choices. **(9 × 2 = 18)**

1. A rose plant can be grown from
   (a) Stem cutting (b) Root (c) Tendril
   (d) Seed (e) Leaf (f) Grafting

2. Identify the animals that belong to polar habitat.
   (a) Walrus (b) Camel (c) Whale
   (d) Penguin (e) Fish (f) Deer

3. Which joint allows movement in all directions?
   (a) Hinge and pivot joint (b) Ball and socket joint
   (c) Hip joint and shoulder joint (d) Wrist and ankle joint
   (e) Backbone and head joint (f) Ankle and hinge joint

4. Identify deficiency disorders from the following.
   (a) Flu (b) Anaemia (c) Chicken pox
   (e) Scurvy (d) Cholera (f) Measles

5. Assume that your friend has sprained his/her ankle. Which of the following would you do?
   (a) Keep his/her ankle on a chair and tilt the position of the ankle.
   (b) Ask him/her not to move the sprained part.
   (c) Ask him/her to get up and run in order to avoid stiffness.
   (d) Wrap an elastic bandage around the sprained area.
   (e) Massage the sprained area with any massage cream and apply heat over it.
   (f) Apply pressure over the sprained area and put some ice packs over it.

6. Identify the uses of oxygen.
   (a) Preparing fertilizers
   (b) Required for respiration
   (c) Increases soil fertility
   (d) Used during photosynthesis
   (e) Required for burning
   (f) For pollination
7. What are the characteristic features of air?
   (a) Exerts pressure
   (b) Does not occupy space
   (c) Weightless
   (d) Mixture of gases
   (e) Pure substance
   (f) Liquid state
8. Which of the following statements are true for a lunar eclipse?
   (a) The moon is covered by the shadow of the Earth
   (b) The Earth is covered by the shadow of the moon
   (c) The sun is covered by the shadow of the moon
   (d) The Earth comes between the sun and the moon
   (e) The moon comes between the sun and the Earth
   (f) The sun comes between the Earth and the moon
9. Which of the following are the layers of the Earth?
   (a) Corona
   (b) Mantle
   (c) Troposphere
   (d) Core
   (e) Crust
   (f) Ozone layer

C Answer the following questions. **(2 × 3 = 6)**

1. What are three types of nerves in the human body?
2. What is solar system? Name the different members of the solar system.

D Answer the following questions in detail. **(4 × 4 = 16)**

1. Briefly explain the processes involved in agriculture.
2. Briefly explain the various types of habitats with their examples.
3. List the symptoms and preventive measures of the following deficiency disorders.
   (a) Beri-beri
   (b) Goitre
4. How will you give first aid to someone bitten by a stray dog?

COMPREHENSIVE ASSESSMENT 2

Chapters 8—14 **(Maximum Marks = 50)**

A Name the following. **(10 × 1 = 10)**

1. The term used for substances that cause pollution
2. The outermost layer of the Earth
3. The region where the Earth's crust tends to be weak and develops a crack
4. One example of sedimentary rock

5. The main constituent of the rock limestone
6. The tiny particles that matter is made up of
7. The category of materials to which an egg belongs (choose from transparent, translucent and opaque)
8. The force that acts on objects thrown upwards
9. The force experienced by like poles of magnets when they are brought closer to each other
10. The energy possessed by an object due to its motion

B In each of the following questions, two choices are correct. Identify the correct choices. **(9 × 2 = 18)**

1. Which of the following causes air pollution?
   (a) Heavy rainfall (b) Smoke from the industries
   (c) Washing clothes in a pond (d) Oil spill
   (e) Using firewood for cooking (f) Loudspeakers

2. Which of the following are igneous rocks?
   (a) Granite (b) Shale (c) Conglomerate
   (d) Marble (e) Pumice (f) Slate

3. The causes of flood are
   (a) Heavy rainfall
   (b) Excess usage of groundwater
   (c) Overflowing water bodies
   (d) Creation of larger number of tributaries for rivers
   (e) Less evaporation of sea water
   (f) Large amount of building constructions

4. What are the applications of air pressure in our daily life?
   (a) Windmill (b) Drinking through a straw (c) Coffee makers
   (d) Thermometer (e) Vacuum cleaner (f) Water treatment

5. Which of these are translucent materials?
   (a) Spectacles (b) Butter paper (c) Cardboard sheet
   (d) Aquarium bowl (e) Frosted glass (f) Wooden block

6. Examples of wheel and axle are
   (a) Doorknob (b) Pulley (c) Axe
   (d) Knife (e) Scissors (f) Bottle opener

7. Force can change the shape of an object. Which of the following are examples of this?
   (a) Blowing air into a balloon
   (b) Goalkeeper stopping a ball in a football match
   (c) Man riding a cycle fast
   (d) Batsman hitting a ball in a cricket match
   (e) Making a model using clay
   (f) Applying brakes in a moving car

8. Rohan and Suman want to measure the length and breadth of one of the floor tiles in their respective houses for comparison. Which of the following units are suitable for measuring them?
   (a) Hand span (b) Metre scale (c) Cubit
   (d) Measuring tape (e) Foot (f) Thumb
9. Which of the following are the units of time measurement?
   (a) Degree Celsius (b) Milligram (c) Days
   (d) Hours (e) Kelvin (f) Metre

C Answer the following questions. **(2 × 3 = 6)**

1. Identify and label the mantle, core and crust in the given diagram.

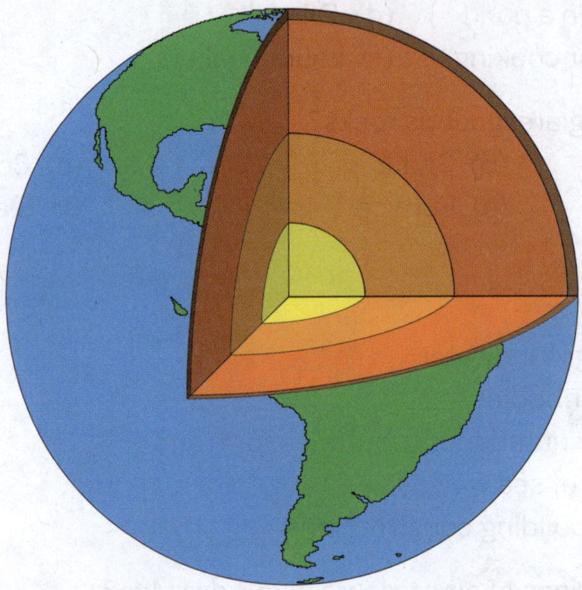

2. What are metamorphic rocks? Briefly explain any two types of metamorphic rocks.

D Answer the following questions in detail. **(4 × 4 = 16)**

1. Briefly explain the causes and effects of land and water pollution.
2. Describe the process of distillation with the help of a labelled diagram.
3. What are the characteristics of a shadow?
4. Explain measurement. Name different quantities of measurement with their standard units and measuring devices.

SOF
SCIENCE OLYMPIAD FOUNDATION

DO NOT OPEN THIS BOOKLET UNTIL ASKED TO DO SO

Roll No.: | | | | | | | | | | | | |

Total Questions: 50 Time: 1 hr.

CLASS 5

SET-A

NSO

18TH SOF NATIONAL SCIENCE OLYMPIAD

Guidelines for the Candidate

1. You will get additional ten minutes to fill up information about yourself on the OMR Sheet, before the start of the exam.
2. The Question Paper comprises three sections :

   Logical Reasoning Section (10 Questions), **Science** Section (35 Questions) and **Achievers Section** (5 Questions)

   Each question in Achievers Section carries 3 marks, whereas all other questions carry one mark each.
3. All questions are compulsory. There is no negative marking. Use of calculator is not permitted.
4. Write your **Name, Roll No.** and **Class** clearly on the **OMR Sheet** and do not forget to sign it.
5. There is only ONE correct answer. Choose only ONE option for an answer.
6. To mark your choice of answers by darkening the circles in the OMR Sheet, use

   HB Pencil or **Blue / Black ball point pen** only. E.g.

Q. 16: Raj goes to school on a bicycle. Which of these safety rules he should follow?

A. He should give signals before taking turns.

B. He should keep to the right of the road.

C. He should cross the road by using zebra crossing.

D. All of these

As the correct answer is option A, you must darken the circle corresponding to option A in the OMR Sheet.

16. ● Ⓑ Ⓒ Ⓓ

7. Rough work should be done in the blank space provided in the booklet.
8. Return the OMR Sheet to the invigilator at the end of the exam.
9. Write your Roll No. on the Question Paper too and take it home for future reference.

1. Which of the following options will continue the series established by the Problem Figures?

Problem Figures

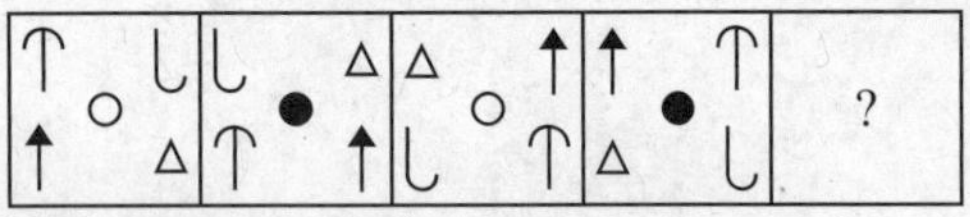

A. 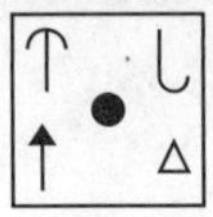
B. 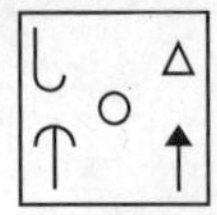
C. 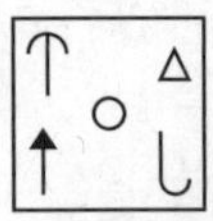
D. 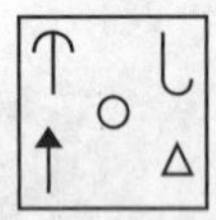

2. Kavya is facing the Theatre. She turns 225° in anti-clockwise direction, where will she be facing now?

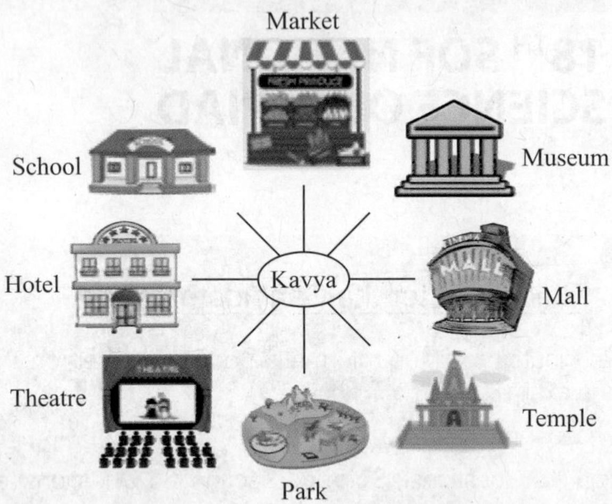

A. School
B. Market
C. Mall
D. Museum

3. If in a certain code language, 'FRIEND' is written as 'IRFDNE', then how will 'MOTHER' be written in the same language?

A. TOMREH
B. TOMEHR
C. REHTOM
D. TOMRHE

4. Which of the following options will complete the pattern in Fig. (X)?

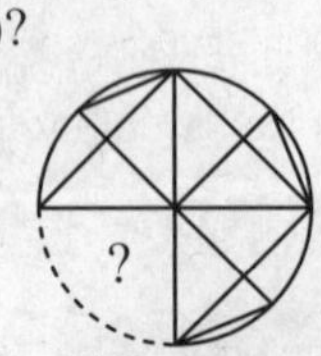

Fig. (X)

A. 
B. 
C. 
D. 

5. Find the missing number, if a certain rule is followed in the given figures.

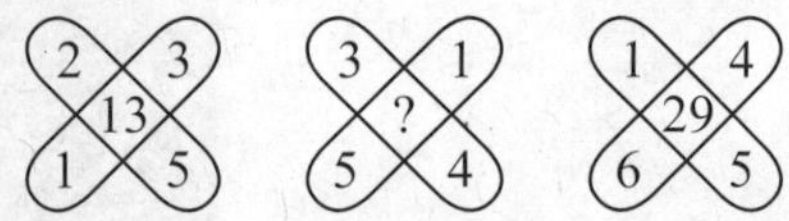

A. 11
B. 15
C. 21
D. 17

6. Select the correct mirror image of Fig. (X).

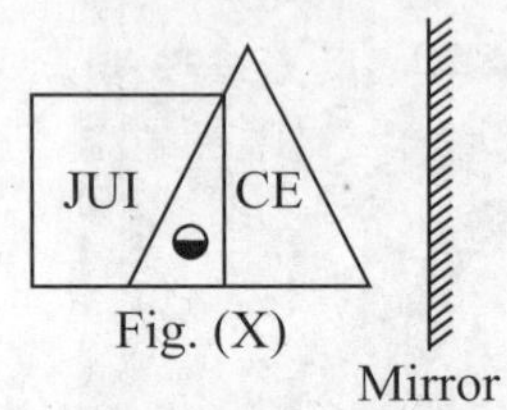

Fig. (X)

A. 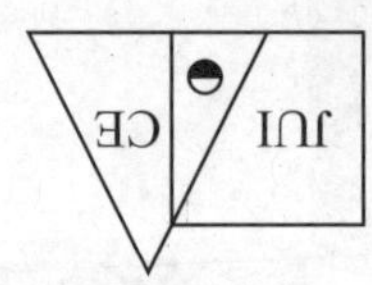
B. 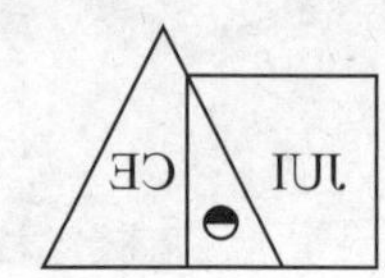
C. 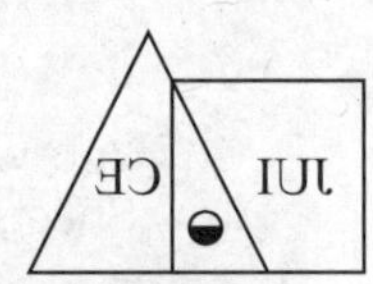
D. 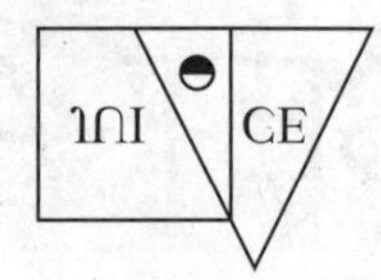

7. Which of the following Venn diagrams best represents the relationship amongst "Players, Men, Champions"?

A. 
B. 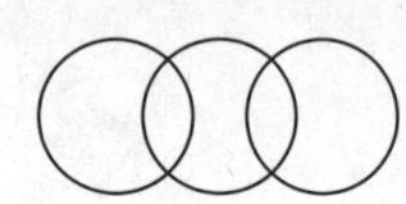
C. 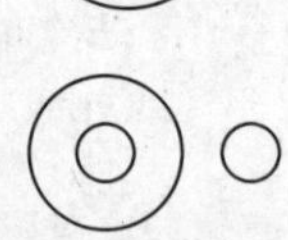
D. 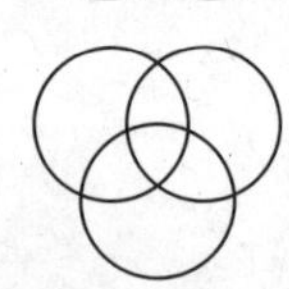

8. Count the number of squares formed in the given figure.

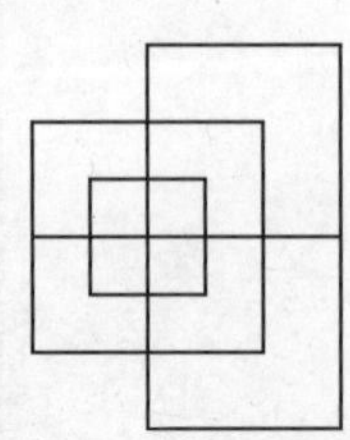

A. 12
B. 14
C. 10
D. None of these

9. Which of the following options is exactly embedded in Fig. (X)?

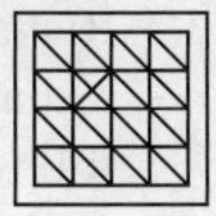

Fig. (X)

A.  B. 

C.  D. 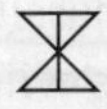

10. Which of the following words CANNOT be formed from the letters of the given word?

CREATIONS

A. REACT
B. ACTION
C. TORN
D. HECTARE

SCIENCE

11. Which of the following statements is incorrect about the Sun?

A. The Sun is a huge mass of very hot gases.
B. The Sun is the only planet that we can see in the day.
C. The distance of the Sun from the Earth is about 150,000,000 km.
D. The Sun is important to the Earth because it provides light energy to green plants to make food.

12. X cannot make its own food and reproduces with the help of tiny structures called spores. Which of the following represents X?

A. Fern
B. Mushroom
C. Bryophyllum
D. Dahlia

13. Amit continued to heat 100 mL of pure water for 15 minutes after it reached boiling point. Which of the following figures shows the result at the end of the experiment?

A. 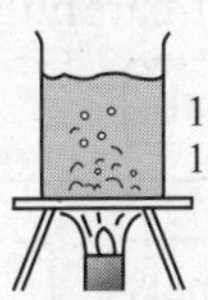 100°C 100 mL

B. 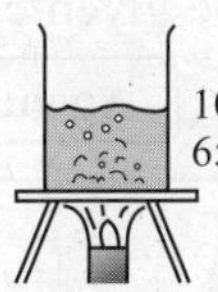 100°C 65 mL

C. 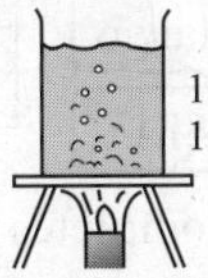 110°C 100 mL

D. 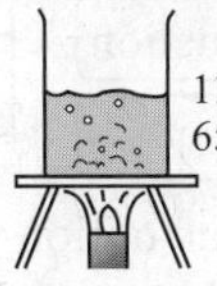 110°C 65 mL

14. The given figure shows the phases of the Moon as seen from Earth on different nights.

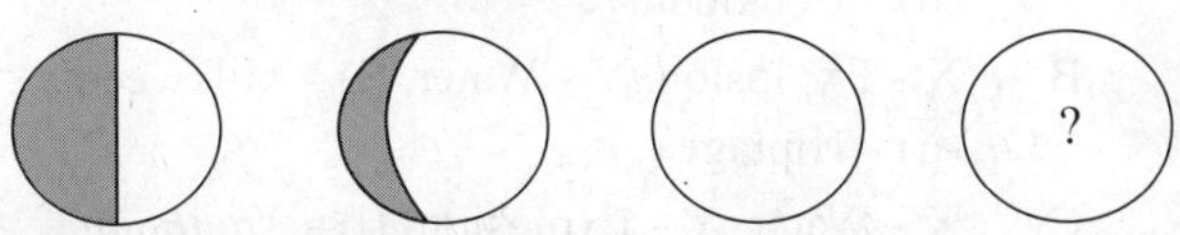

December 21 December 26 December 28 December 30

Which one of the following figures shows the correct phase of the Moon on December 30?

A. 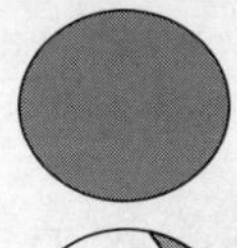 B. 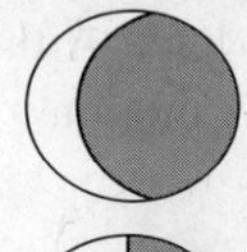

C. 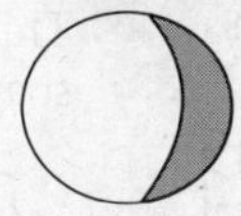 D. 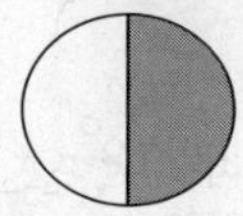

15. Study the given classification chart.

Materials

Group X	Group Y
Styrofoam Wooden chopstick Rubber duck	Metal coin Ceramic bowl Glass marble

Which of the following are the most suitable headings for the groups X and Y?

	Group X	Group Y
A.	Man-made	Natural
B.	Flexible	Stiff
C.	Opaque	Transparent
D.	Floats on water	Sinks in water

16. Ritika was asked to draw human digestive system. The diagram drawn by her is shown here. If we had a digestive system, as shown here, what would be the consequence?

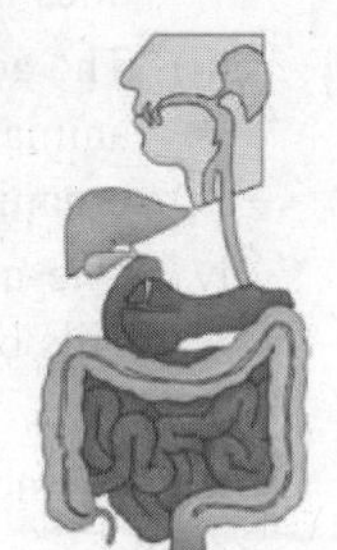

A. The food will not be churned well and as a result, digestion will be slower.
B. Bile will not be produced.
C. Bacteria present in the food will not be killed.
D. Both A and C

17. Soil pollution can be reduced by increasing the use of ______.

A. Biodegradable and recyclable materials
B. Use-and-throw articles
C. Electronic and plastic products
D. Forest products without reforestation

18. Mohan is very sensitive to green chillies. Whenever he eats them, he gets rashes on his body. Mohan has a disorder called ______.

A. Measles B. Allergy
C. Anaemia D. Cholera

19. The organism 'P' migrates to a new place, reproduces and then dies. Which of these can be 'P'?

A. Salmon B. Polar bear
C. Arctic tern D. Siberian crane

20. Aradhya dips a can into a container of soapy water as shown here. A thin film of soapy water was formed over its mouth.

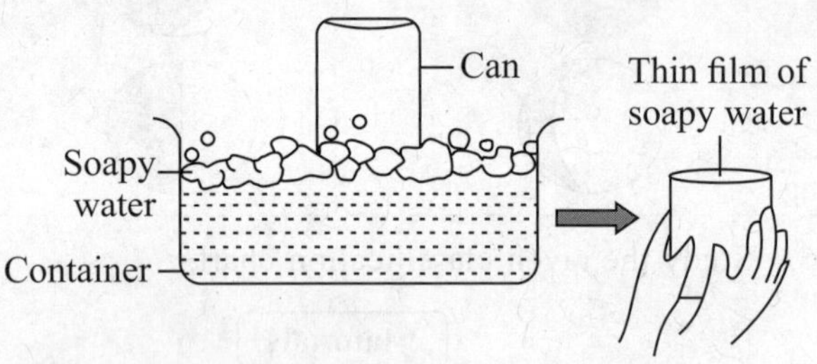

Aradhya holds the can tightly in her hands for a while. What do you think will happen?

A. Nothing will happen.
B. A bubble will form out of the mouth of the can.
C. The film will sink into the can.
D. The film will pop.

21. Which of the following statements show correct differences between an adult mosquito and its young one?

(i) The adult mosquito lives on land but its young one lives in water.
(ii) The adult mosquito moults but its young one does not.
(iii) The adult female mosquito feeds on the blood of animals but its young one feeds on microscopic organisms.
(iv) The adult mosquito breathes in air through its gills but its young one breathes through a special tube.

A. (i) and (ii) only B. (ii) and (iv) only
C. (i) and (iii) only D. (i), (iii) and (iv) only

22. Which of the following sets of gases contributes to global warming?

A. Sulfur dioxide, Nitrogen, Carbon dioxide and Ozone
B. Oxygen, Carbon dioxide, Ozone and Methane
C. CFC, Oxygen, Nitrogen and Carbon dioxide
D. Methane, Carbon dioxide, Nitrous oxide and Water vapour

23. Rishi is a contracter who is clearing a forest patch to build a sports complex in its place. Which of the following would possibly be the consequence of this activity?

(i) Some animals will become extinct.
(ii) Some possible sources of natural drugs to cure diseases will be lost.
(iii) The problem of global warming will reduce.
(iv) Water table will replenish.

A. (i) and (ii) only
B. (i), (ii) and (iii) only
C. (ii) and (iv) only
D. (i), (ii), (iii) and (iv)

24. The material used to make which of the following items can also be used to make a box for transporting cooked food items in order to keep them hot?

A. 
Metal fork

B. 
Wooden spoon

C. 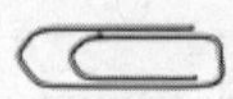
Paper clip

D. 
Frying pan

25. Which of the following statements is correct?

A. The mass movement of animals from one place to another is called hibernation.
B. Most materials from living resources are biodegradable.
C. Bones store essential body minerals like calcium and iron.
D. Rice, maize and hemp are rabi crops.

26. Read the given classification chart carefully.

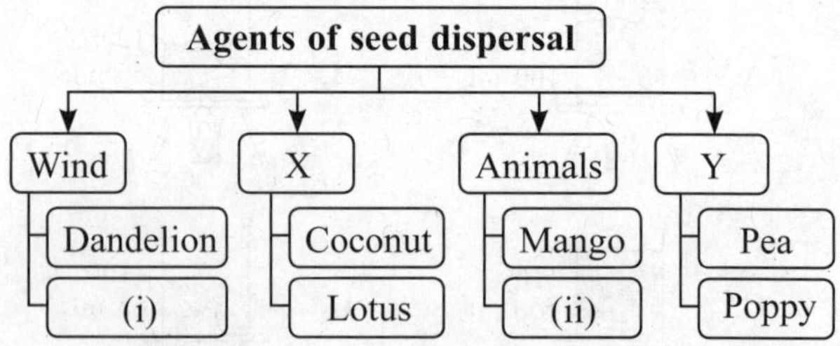

Select the option which correctly completes the given classification chart.

A. X - Water, Y - Explosion, (i) - Maple (ii) - Cocklebur
B. X - Explosion, Y - Water, (i) - Milkweed (ii) - Hiptage
C. X - Water, Y - Explosion, (i) - *Xanthium* (ii) - Cherry
D. X - Explosion, Y - Water, (i) - Apple, (ii) - Cotton

27. Which of the following options correctly represents the conditions present on land and water when a sea breeze flows?

	On land	On water
A.	High pressure	Low pressure
B.	High temperature	Low temperature
C.	Low pressure	Low pressure
D.	Low temperature	High temperature

28. When night becomes very cold, X freezes to form Y. What can be X and Y here?

	X	Y
A.	Hail	Snow
B.	Dew	Frost
C.	Frost	Dew
D.	Snow	Hail

29. In a newly cleared area three plants X, Y and Z were planted at the boundary. Plant X spread over the entire area with time, while plants Y and Z were found limited in the area near the boundary. Select the correct option regarding it.

A. Seeds of plant X are dispersed by explosion.
B. Seeds of plant Y are dispersed by wind.
C. Seeds of plant Z are dispersed by animals.
D. None of these

30. The amount of sugar dissolved in three beakers P, Q and R is represented in the given graph.

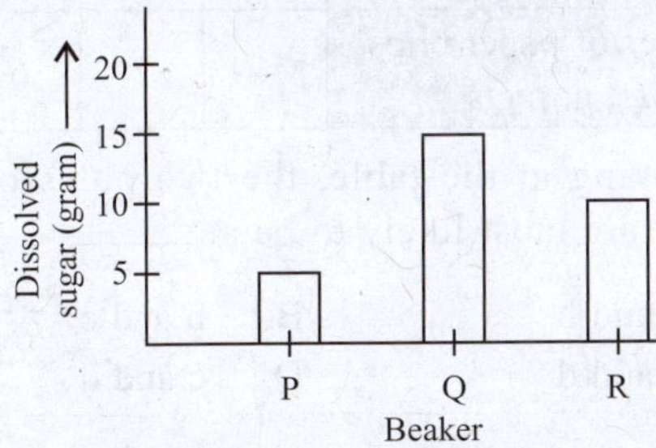

Which of the following inferences can be made from this?

A. If all the beakers have equal amount of water, then beaker Q is at the highest temperature.
B. If all the beakers have equal temperature of water, then beaker P has the lowest amount of water.
C. If all the beakers have equal amount of water, then the beaker P is at the lowest temperature.
D. All of these

31. Select the incorrect statement among the following.

A. The revolution period of a planet depends upon its distance from the Sun.
B. The rotation period of planet depends upon its distance from the Sun.
C. The duration of a day on a planet depends upon its rotatory speed.
D. The duration of a year on a planet depends upon its revolution period.

32. What will happen when a flask of coloured water is placed in a container of hot water for some time, as shown here?

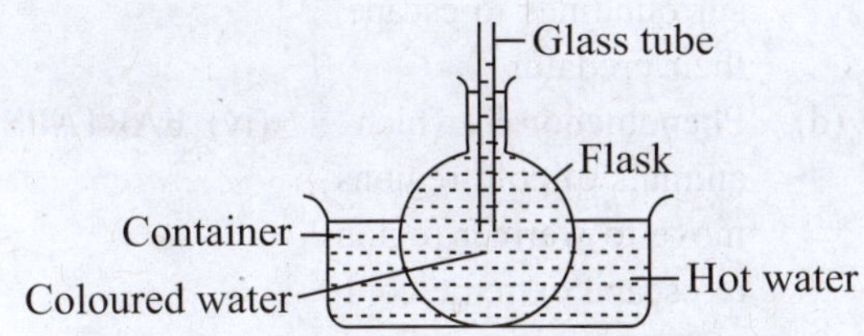

A. The water in the container will overflow.
B. The water level in the glass tube will fall.
C. The water level in the glass tube will rise.
D. The water in the container will become coloured.

33. Study the given Venn diagram and select the position where Moon can be placed in this diagram.

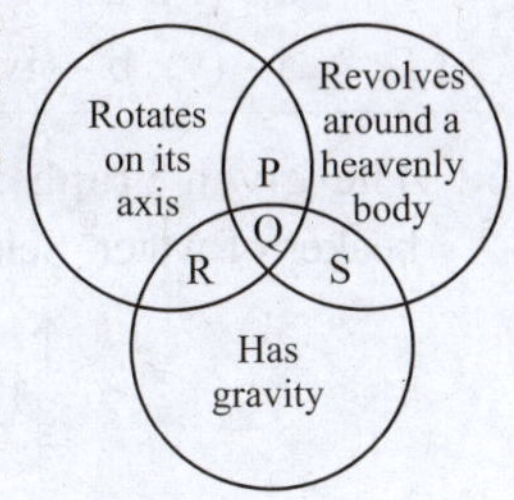

A. P
B. Q
C. R
D. S

34. Select the option that correctly represents the arrangement of Load (L), Fulcrum (F) and Effort (E) in a second class lever.

A. 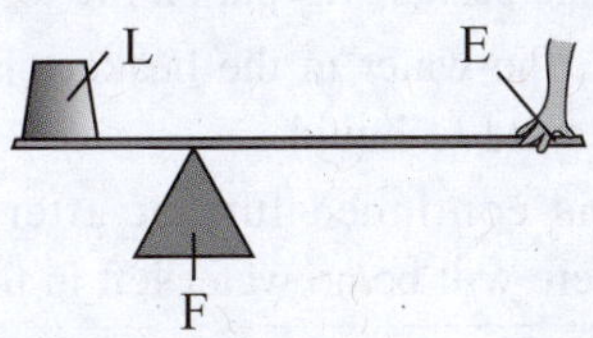

B. 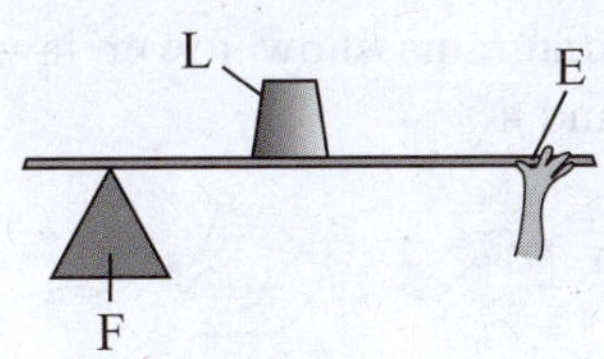

C. 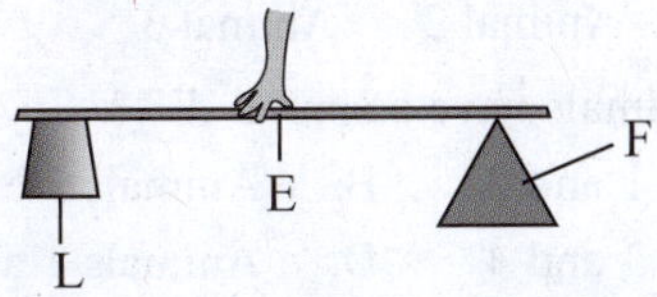

D. 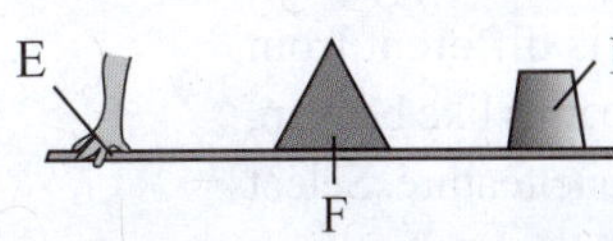

35. Match the column I (characteristics or phenomenon) with column II (containing scrambled letters) and select the correct option from the given codes.

	Column I		Column II
(a)	Animal that eats both plants and animals	(i)	GMOANIRTI
(b)	Animal that depends on other living animals for food	(ii)	OFULCMAEGA
(c)	Phenomenon in which animals merge with their surroundings to escape their predator	(iii)	AEARLBOR
(d)	Phenomenon in which animals of cold regions move to warmer regions to escape extreme cold winters	(iv)	EARTAPSI
(e)	Animal that spends most time on trees	(v)	NVMOIOER

A. a - (iv), b - (iii), c - (v), d - (ii), e - (i)
B. a - (i), b - (ii), c - (iii), d - (iv), e - (v)
C. a - (ii), b - (i), c - (iv), d - (v), e - (iii)
D. a - (v), b - (iv), c - (ii), d - (i), e - (iii)

36. The given graph shows temperature changes in a beaker of water. Select the incorrect option regarding it.

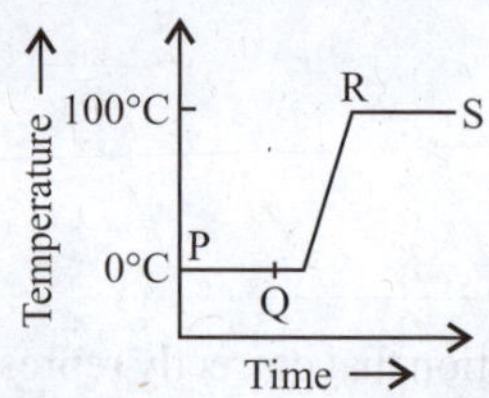

A. At point P, the beaker was in refrigerator.
B. At point Q, the beaker was put on fire to be heated.
C. At point R, the water in the beaker changed its phase from solid to liquid.
D. If heating is continued further after point S, gradually there will be no water left in the beaker.

37. The following diagrams show lower jaw of four animals 1, 2, 3 and 4.

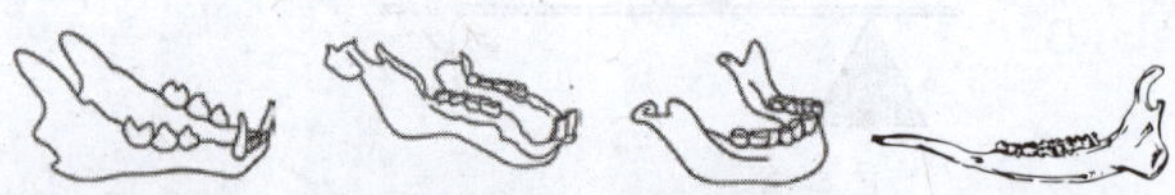

Animal 1 Animal 2 Animal 3 Animal 4

Which two animals have a similar diet?
A. Animals 1 and 2 B. Animals 2 and 4
C. Animals 3 and 4 D. Animals 1 and 3

38. The given plant is different from the trees found near us like banyan, neem etc., in a basic feature. Select the correct option regarding it.

A. This plant is found in water.
B. This plant obtains its nutrition from other plants.
C. This plant obtains its nutrition completely from insects.
D. This plant has special leaves which snap shut when sense touch.

39. Study the given classification chart. Which of the bones is classified incorrectly?

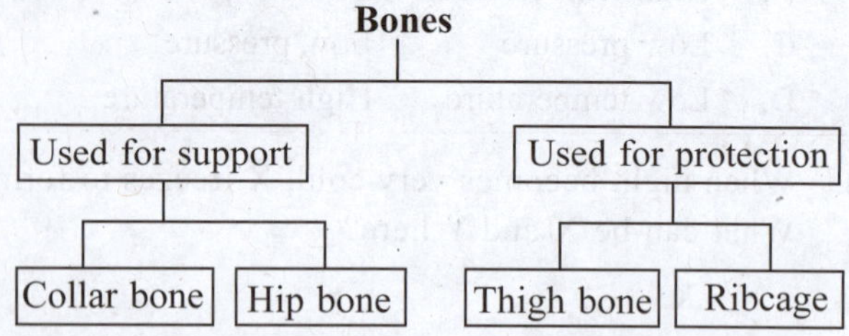

A. Thigh bone B. Hip bone
C. Collar bone D. Ribcage

40. Clothes should be cared properly for their durability and hygiene. Select the correct option regarding care of clothes.
A. Moth balls should be placed in the cupboard but they should not touch the clothes.
B. Silk clothes should be washed with mild detergents.
C. Woollen clothes should be wrung very tightly before hanging for drying.
D. Both A and B

41. The given table shows four parts of a bar magnet and the number of paper clips each part attracts.

Parts of a magnet	a	b	c	d
Number of paper clips each part attracts	4	7	6	3

By looking at the table, the two poles of the bar magnet are most likely to be at ______.

A. a and b B. b and c
C. a and d D. c and d

42. Ritu has a toy puppy that can bark, jump, wag tail and move. Its eyes shine when patted. It is operated by a battery. Which of the following energy conversions do not take place in this puppy?
A. Chemical energy → Kinetic energy
B. Chemical energy → Sound energy
C. Chemical energy → Light energy
D. Chemical energy → Magnetic energy

43. Study the two life cycles carefully.

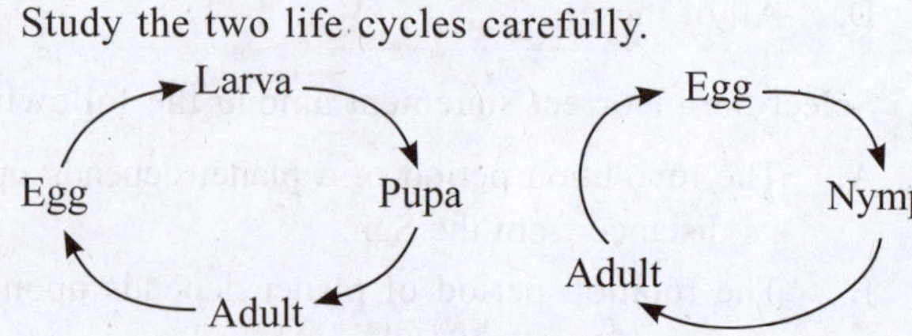

Which of the following statements is true about the life cycles given here?

A. In both the life cycles, newly hatched young ones do not resemble the adults.

B. In both the life cycles, all stages occur in water.

C. In both the life cycles, the newly hatched young ones resemble the adults.

D. In both the life cycles, the adults moult many times.

44. A freshwater river supports a large number of fish in its water. A new chemical factory is being established near this river. Select the graph that represents the most likely changes in the fish population in river in future. In graphs, Q represents the time point when factory is established.

A. 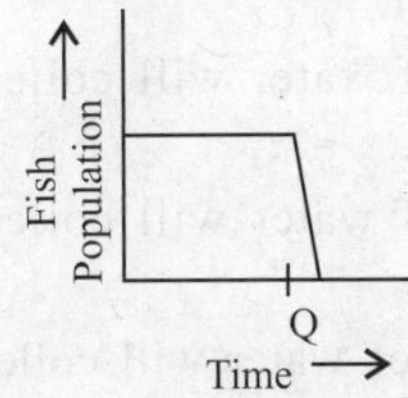

B. 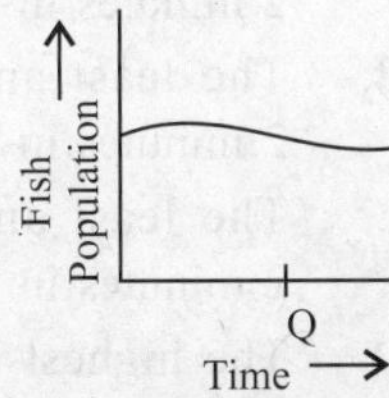

C. 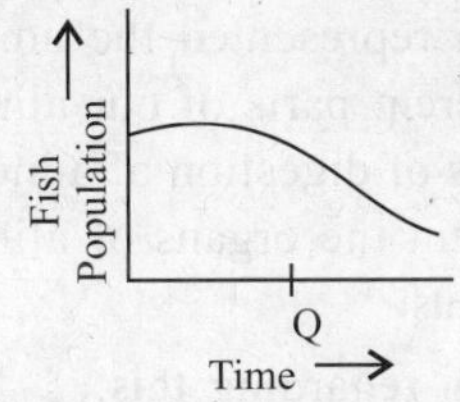

D. 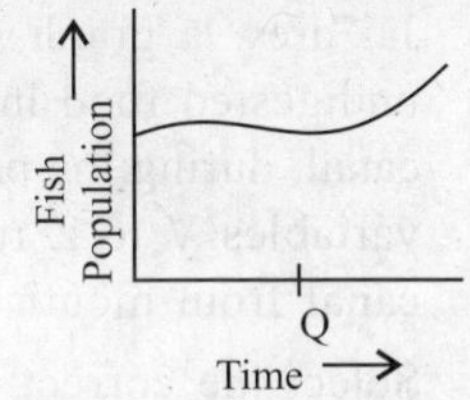

45. Figure P shows an object which is made of three different materials. Figure Q shows its shadow on a screen. Study these figures and select the option that correctly identifies any two of these materials.

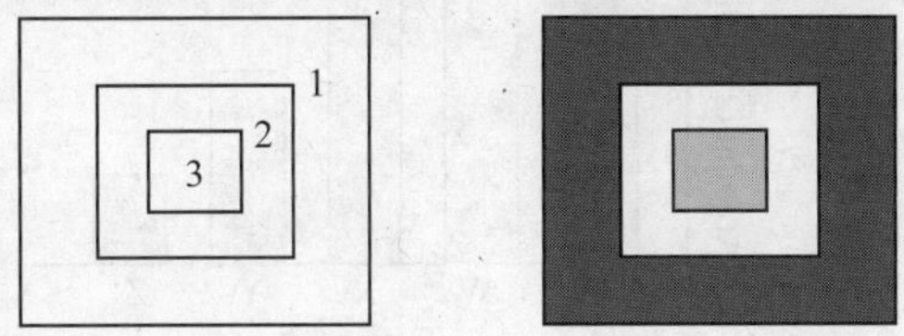

P Q

A. 1 - Tissue paper, 2 - Cardboard

B. 2 - Cardboard, 3 - Cellophane sheet

C. 1 - Cardboard, 3 - Tracing paper

D. 2 - Frosted glass, 3 - Cardboard

ACHIEVERS SECTION

Direction : Refer to the given word grid and answer the questions 46 and 47.

46. Which of the following materials is not hidden in this word grid?

P	Q	S	M	D	T	V	R	L	J
R	A	S	S	I	L	M	U	P	K
I	P	Q	N	A	G	H	B	R	E
E	R	O	A	M	F	K	B	O	F
C	O	T	T	O	N	I	E	T	J
L	R	A	I	N	K	J	R	E	P
F	M	P	B	D	G	L	A	S	S

A. The hardest material in the world.

B. Material sourced from plants that is used to fill pillows.

C. Material used to make transparent and delicate decoratives.

D. Clothing material that is obtained by killing young one of an animal.

47. A material having a six letter name starting with R is hidden in this word grid. Which of the following objects cannot be made using this material?

A. Truck tyres B. Bulb

C. Eraser D. Hose pipes

48. Padma conducted an experiment as shown below by using different materials.

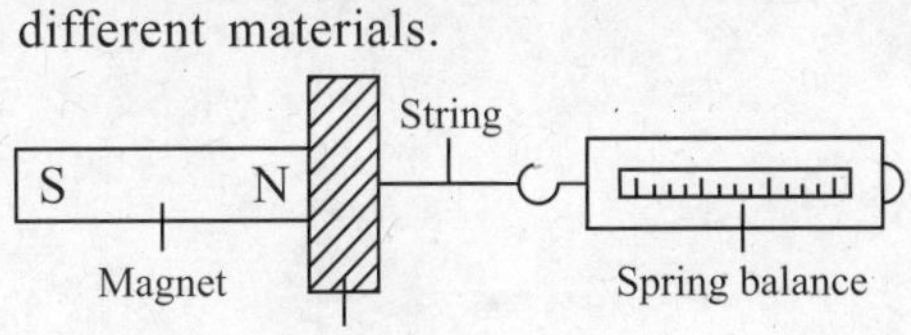

Padma pulled on the spring balance until the material was separated from the magnet. The readings on the spring balance were recorded as shown in the table below.

Material	Reading on the Spring Balance (g)
Steel	68
Soft iron	82
Nickel	38
Cobalt	48

What conclusions could Padma make from her experiment?

(i) Soft iron needs the largest force in order to be separated from a magnet.

(ii) Nickel is most easily magnetised while soft iron is least easily magnetised.

(iii) Nickel and cobalt are non-magnetic materials.

(iv) Steel makes the strongest permanent magnet.

A. (i) only B. (iii) and (iv) only

C. (i), (ii) and (iii) only D. (iv) only

49. Jai drew a graph which represented the amount of undigested food in different parts of our alimentary canal, during the process of digestion of a meal. The variables V to Z represent the organs of alimentary canal from mouth to anus.

Select the correct option regarding this.

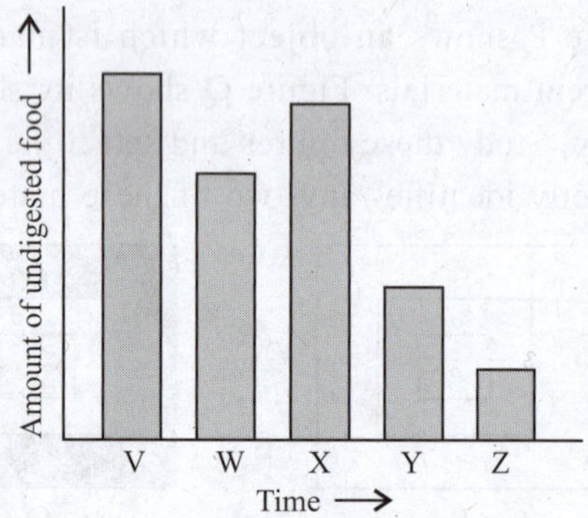

A. The amount of undigested food should be zero at *Z* because digestion is completed before the food reaches anus.

B. The amount of undigested food at *X* and *W* should be interchanged because there is no digestion in oesophagus.

C. A decrease in the amount of undigested food should be first observed in *Y* only because intestine is the primary site of digestion.

D. Both A and B

50. Rohit prepared three set-ups as shown here. He then poured 100 mL of water into each flask.

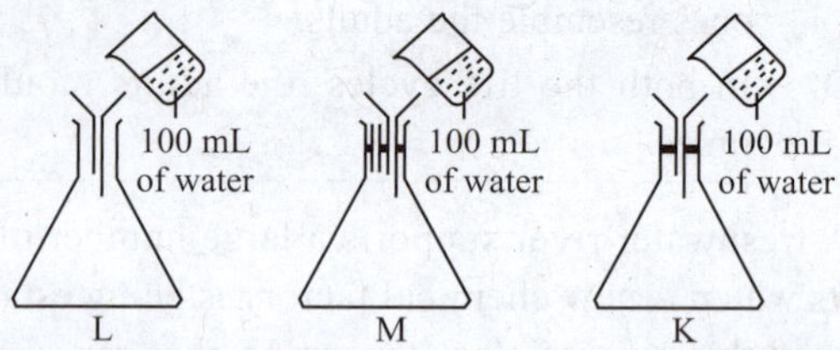

Which of the following assumptions holds true for the given experiment?

A. The highest amount of water will collect after 2 minutes in beaker M.

B. The least amount of water will collect after 2 minutes in beaker K.

C. The least amount of water will collect after 2 minutes in beaker L.

D. The highest amount of water will collect after 2 minutes in beaker K.

SPACE FOR ROUGH WORK

The Story of Ozone

What is the ozone layer? Where is it? I do not see anything from here.
It is high up in the atmosphere; that is why we cannot see it. It prevents harmful rays of the sun from reaching the Earth.
Then, is it like a blanket?
Yes, but now there is a hole in it!
Really? Who made it?
The humans and the pollution they create.
Is the hole a real thing? Oh my god! Now what do we do?
Yes, my children; it is true. Harmful rays from the sun can now come to the Earth through the hole.

What will happen now, Mamma?
Don't worry, human beings are learning from their mistakes.
Mamma, look, some school children are passing by. What are they doing? And what are they holding in their hands?
Read my children, they are creating awareness on reducing pollution and saving the Earth.
Do not cut trees; instead, plant new ones.
Say no to plastics! Carry your own cloth bag.
Reduce the smoke from vehicles; get your vehicle tested regularly.
Do not burn waste, reuse or recycle them.
Do not burn crackers; instead, light a lamp.
Keep your environment clean. Do not litter.
Do not pollute water sources.
Do not waste water. Harvest rainwater.

Notes

Notes

Notes